T0137889

Lecture Notes in Computer Science 14314

The series Lecture Notes in Computer Science (LNCS), including its subseries Lecture Notes in Artificial Intelligence (LNAI) and Lecture Notes in Bioinformatics (LNBI), has established itself as a medium for the publication of new developments in computer science and information technology research, teaching, and education.

LNCS enjoys close cooperation with the computer science R & D community, the series counts many renowned academics among its volume editors and paper authors, and collaborates with prestigious societies. Its mission is to serve this international community by providing an invaluable service, mainly focused on the publication of conference and workshop proceedings and postproceedings. LNCS commenced publication in 1973.

Binod Bhattarai · Sharib Ali · Anita Rau ·
Anh Nguyen · Ana Namburete ·
Razvan Caramalau · Danail Stoyanov
Editors

Data Engineering in Medical Imaging

First MICCAI Workshop, DEMI 2023
Held in Conjunction with MICCAI 2023
Vancouver, BC, Canada, October 8, 2023
Proceedings

Editors
Binod Bhattarai
University of Aberdeen
Aberdeen, UK

Sharib Ali
University of Leeds
Leeds, UK

Anita Rau
Stanford University
Stanford, CA, USA

Anh Nguyen
University of Liverpool
Liverpool, UK

Ana Namburete
University of Oxford
Oxford, UK

Razvan Caramalau
University College London
London, UK

Danail Stoyanov
University College London
London, UK

ISSN 0302-9743 ISSN 1611-3349 (electronic)
Lecture Notes in Computer Science
ISBN 978-3-031-44991-8 ISBN 978-3-031-44992-5 (eBook)
https://doi.org/10.1007/978-3-031-44992-5

This Springer imprint is published by the registered company Springer Nature Switzerland AG
The registered company address is: Gewerbestrasse 11, 6330 Cham, Switzerland

Paper in this product is recyclable.

Preface

DEMI 2023 was the "1st International Workshop on Data Engineering in Medical Imaging", organized as a satellite event of the 26th International Conference on Medical Image Computing and Computer Assisted Intervention (MICCAI 2023) in Vancouver, Canada. Data engineering plays a vital role in advancing medical imaging research, where limited data availability poses a significant challenge. To tackle this issue, the medical imaging community has adopted various techniques, including active learning, label and data augmentation, self-supervision, and synthetic data generation. However, the potential of these methods has yet to be fully leveraged. Data augmentation, for instance, is often randomly chosen based on intuition. Yet previous work has shown that it is crucial to jointly optimize the augmentations' complexity and affinity, i.e., how much the augmentation shifts the decision boundary of the clean baseline model. Other studies suggest that not every synthetic example improves the model's generalizability, with some even hurting performance if not reasonably chosen. Similarly, the effect of self-supervised pre-training methods on downstream tasks generally depends on the overlap between the pretext task and the downstream tasks. For instance, a model trained to predict rotation angles will not be effective in rotationally invariant organs. Thus, DEMI aims to invite researchers to submit their work in the field of medical imaging around the central theme of data engineering in various topics such as data and label augmentation, active learing and active synthesis, federated learning, multimodal learning, self-supervised learning and large-scale data management and data quality assessment.

The DEMI 2023 proceedings contain 11 high-quality papers of 9 to 15 pages preselected through a rigorous peer review process. All submissions were peer-reviewed through a double-blind process by on average three members of the scientific review committee, comprising 16 experts in the field of medical imaging. The accepted manuscripts cover various medical image analysis methods and applications. In addition to the papers presented in this LNCS volume, the workshop hosted two keynote presentations from world-renowned experts: James Zhou (Stanford University) and Qi Dou (Chinese University of Hong Kong). We wish to thank all the DEMI 2023 authors for their participation and the members of the scientific review committee for their feedback and commitment to the workshop. We are very grateful to our sponsor FogSphere.

October 2023

Binod Bhattarai
Sharib Ali
Anita Rau
Anh Nguyen
Ana Namburete
Razvan Caramalau
Danail Stoyanov

Organization

Program Committee Chairs

Binod Bhattarai	University of Aberdeen, UK
Sharib Ali	University of Leeds, UK
Anita Rau	Stanford University, USA
Anh Nguyen	University of Liverpool, UK
Ana Namburete	University of Oxford, UK
Razvan Caramalau	University College London, UK
Danail Stoyanov	University College London, UK

Technical Program Committee

Adrian Krenzer	Julius Maximilian University of Würzburg, Germany
Annika Reinke	German Cancer Research Center, Germany
Bidur Khanal	Rochester Institute of Technology, USA
Chloe He	University College London, UK
Francisco Vasconcelos	University College London, UK
Gilberto Ochoa-Ruiz	Tec de Monterrey, Mexico
Jialang Xu	University College London, UK
Jingxuan Kang	University of Liverpool, UK
Josiah Aklilu	Stanford University, USA
Mariia Dmitrieva	Queen Square Analytics, UK
Michal Byra	RIKEN Center for Brain Science, Japan
Prashnna Gyawali	West Virginia University, USA
Rumeysa Bodur	Imperial College London, USA
Stefan Denner	German Cancer Research Center, Germany
Tianhong Dai	University of Aberdeen, UK
Tudor Jianu	University of Liverpool, UK
Ziyang Wang	Oxford University, UK

Contents

Weakly Supervised Medical Image Segmentation Through Dense Combinations of Dense Pseudo-Labels

Ziyang Wang(✉) and Irina Voiculescu

Department of Computer Science, University of Oxford, Oxford, UK
{ziyang.wang,irina}@cs.ox.ac.uk

Abstract. Annotating a large amount of medical imaging data thoroughly for training purposes can be expensive, particularly for medical image segmentation tasks. Instead, obtaining less precise scribble–like annotations is more feasible for clinicians. In this context, training semantic segmentation networks with limited-signal supervision remains a technical challenge. We present an innovative scribble-supervised approach to image segmentation via densely combining dense pseudo-labels which consists of groups of CNN– and ViT–based segmentation networks. A simple yet efficient dense collaboration scheme called *Collaborative Hybrid Networks* (CHNets) ensembles dense pseudo–labels to expand the dataset such that it mimics full-signal supervision. Additionally, internal consistency and external consistency training of the collaborating networks are proposed, so as to ensure that each network is beneficial to the others. This results in a significant overall improvement. Our experiments on a public MRI benchmark dataset demonstrate that our proposed approach outperforms other weakly-supervised methods on various metrics. The source code of CHNets, ten baseline methods, and dataset are available at https://github.com/ziyangwang007/CV-WSL-MIS.

Keywords: Weakly-Supervised Learning · Convolution · Vision Transformer · Image Segmentation · Pseudo Labels

1 Introduction

Recent studies of Convolutional Neural Networks (CNN) and self-attention-based Vision Transformers (ViT) have shown exhilarating performance in medical image analysis [6,20,24,28,29]. Most of the recent studies reported state-of-the-art achievement, however, relying on a large-scale set with high-quality pixel-level annotations for training [6,15,16]. To tackle the expensive cost of annotation for segmentation purpose, existing works present network training with Semi-Supervised Learning(SSL) and Weakly-Supervised Learning(WSL) [8,17,18,26,27]. SSL medical image segmentation involves training a model with a few pixel-level labeled data and a large amount of raw data. As an alternative simple way for clinicians to annotate data, this paper presents a

B. Bhattarai et al. (Eds.): DEMI 2023, LNCS 14314, pp. 1–10, 2023.
https://doi.org/10.1007/978-3-031-44992-5_1

WSL scribble-based approach to train CNN-based and ViT-based segmentation networks simultaneously and collaboratively.

WSL for segmentation is normally proposed to leverage sparse annotations including bounding boxes, points, text, and scribbles for model training [13,19,31]. Among them, scribble annotation, seen in Fig. 1, is one of the practical and convenient forms of clinicians labeling. The lack of sufficient information signal, however, remains challenging for medical image semantic segmentation, especially for the classification of pixels on the boundaries of regions of interest. The current SSL and WSL approaches mostly utilize partial-supervision losses to initialize the model and leverage the prior assumption to expand data. By doing so, the inference of the model can be used to expand scribbles and regenerate dense pixel-level pseudo-labels. ScribbleSup [13] proposed a graphical model that propagates feature information from scribbles to unlabeled pixels with a unique loss for model training. Conditional random field [4] was explored to refine the segmentation inference via random walker in an iterative two-step stage to train segmentation model [7]. Scribble2Label [12] was introduced to strengthen pseudo-labeling with a novel label filtering with EMA [22] to generate more reliable pseudo-labels during training. Some works introduce generating a virtual training set by MixUp [30]. For example, CycleMix [31] introduced integrating mix augmentation and regularization of supervision from consistency for scribble-supervised segmentation. Inspired by generative adversarial training, other works propose to encourage high-quality pseudo-labels by introducing an additional model for evaluation. AAG [23], Adversarial Attention Gate, explored adversarial training for the segmentation model with multi-scale attention gates. Adversarial training requires additional computational costs with challenging training settings for additional models.

Recent studies on SSL and WSL have argued that the consistency of pseudo-labels under feature- and network-perturbation is essential for segmentation performance, as consistency-aware training. Triple-view learning [27] introduced three different segmentation networks to iteratively generate pseudo-labels to help each model. Cross teaching [17] further explored Cross Pseudo Supervision [5] between CNN and ViT for SSL. Mix pseudo supervision [18] was then proposed as a data perturbation technique for pseudo label generation, achieving state-of-the-art performance for scribble-supervised MRI cardiac segmentation.

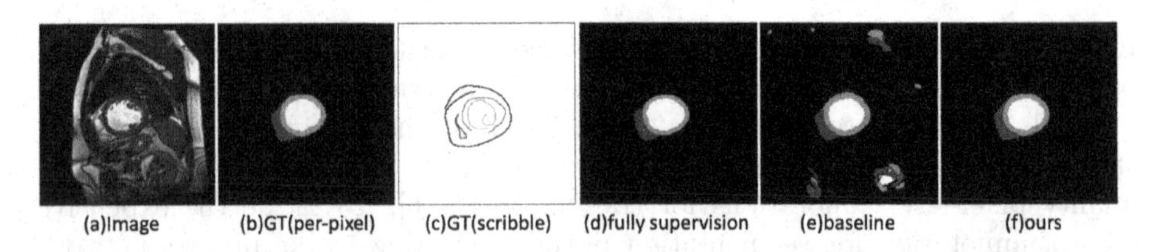

Fig. 1. The Illustration of a Multi–Class Scribble–Supervised Segmentation. (a) Input sagittal left–facing MRI, (b) ground truth dense labels, (c) scribble annotations, (d) segmentation inference by fully supervised UNet, (e) segmentation inference by scribble–supervised UNet, (f) segmentation inference by scribble–supervised CHNets.

Building on the recent advancements in network architecture engineering and consistency-aware training with WSL, we propose Collaborative Hybrid Networks (CHNets) for learning from scribbles. CHNets comprises two feature learning networks: a CNN-based encoder-decoder UNet [20], and a Swin-Transformer-based UNet-style network called SwinUNet [3], which directly replaces the pure CNN layers of UNet to pure self-attention layers of ViT. Our approach aims to facilitate simultaneous and collaborative learning between the two networks by introducing an iterative labeling-ensemble scheme to generate dense pseudo-labels and retrain networks via external-consistency supervision. Additionally, we employ a self-ensemble technique on each network separately under internal-consistency supervision to boost their performance further. Through this dual consistency supervision mechanism, CHNets fully exploits and strengthens the two segmentation networks to produce dense pixel-level inference. We evaluate CHNets on a public scribble-supervised MRI cardiac dataset [2], and our experimental results demonstrate that our approach outperforms other existing WSL methods [3,8,11–14,20,21] on various evaluation metrics.

2 Approach

The proposed CHNets is sketched in Fig. 2, consisting of a CNN-based UNet $f_{\text{cnn}} : \boldsymbol{X} \mapsto \boldsymbol{Y}_{\text{cnn}}$, and ViT-based SwinUNet as $f_{\text{vit}} : \boldsymbol{X} \mapsto \boldsymbol{Y}_{\text{vit}}$, where $\boldsymbol{X} \in \mathbb{R}^{h \times w}$, $\boldsymbol{Y} \in [0, 1, 2, 3]^{h \times w}$ represents a 2D input image, corresponding inference by UNet and SwinUNet, respectively. We denote a batch of scribble-annotated data of training set as $(\boldsymbol{X}, \boldsymbol{Y}_{\text{scrib}}) \in \mathbf{T}_{train}$ where $\boldsymbol{Y}_{\text{scrib}} \in [0, 1, 2, 3, None]^{h \times w}$ ($None$ indicates no annotation information on the corresponding pixels), and densely-annotated data of the test set (e.g. in the form of black and white masks) as $(\boldsymbol{X}, \boldsymbol{Y}_{\text{gt}}) \in \mathbf{T}_{test}$ where $\boldsymbol{Y}_{\text{gt}} \in [0, 1, 2, 3]^{h \times w}$.

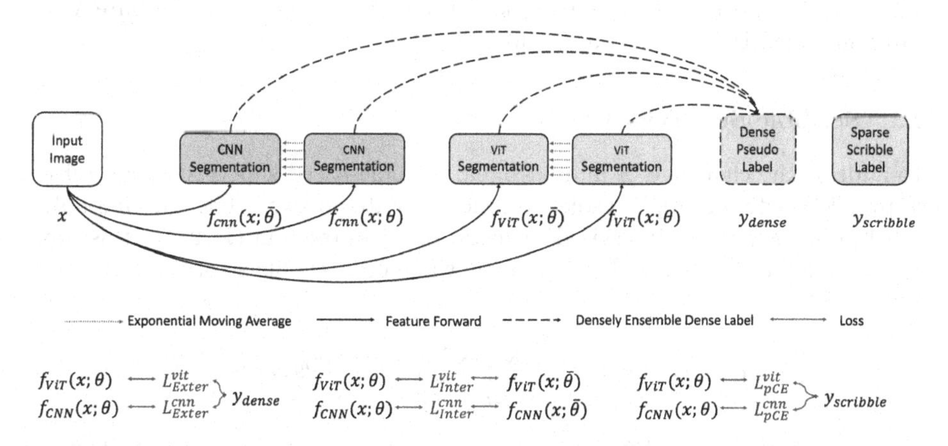

Fig. 2. The Framework of Collaborative Hybrid Networks for Medical Image Segmentation Under Scribble limited–signal Supervision and Dense Pseudo Label full–signal Supervision. It consists of dual ViT–based & CNN–based segmentation networks. Each of the losses are shown.

2.1 Training Objective

The training of CHNets is done in an end-to-end manner and is briefly illustrated in Fig. 2. In general, the whole framework optimizes the hybrid networks using the sum of different categories of losses, formulated as:

$$\mathcal{L} = \underbrace{\mathcal{L}_{\text{pCE}}^{cnn} + \mathcal{L}_{\text{pCE}}^{vit}}_{Scribble-Supervision} + \underbrace{\lambda_1(\mathcal{L}_{\text{inter}}^{cnn} + \mathcal{L}_{\text{inter}}^{vit})}_{Internal-Consistency} + \underbrace{\lambda_2(\mathcal{L}_{\text{exter}}^{cnn} + \mathcal{L}_{\text{exter}}^{vit})}_{External-Consistency} \quad (1)$$

where the scribble-supervision loss, internal-consistency loss, and external- consistency loss are indicated as $\mathcal{L}_{\text{pCE}}$, $\mathcal{L}_{\text{inter}}$, $\mathcal{L}_{\text{exter}}$, respectively. Depending on the architecture of the segmentation networks used (in this case CNN-based or ViT-based networks), losses are also qualified as $\mathcal{L}^{cnn}$, or $\mathcal{L}^{vit}$. The λ is a ramp-up function which enforces the training gradually to move from limited-signal scribble-supervised learning to dense pseudo-full-signal supervised learning [25].

2.2 CNN– And ViT–Based Networks

Motivated by the success of the legendary UNet and recent advancements in Vision Transformers (ViT), we intuitively designed a hybrid network architecture for single segmentation task. Recent strategies for training multi-networks with limited signal include Multi-View Learning [27] feeding with different augmented data, Co-Teaching [9] with noisy labels, and Cross Supervision [5] with different parameter initializations. All of these approaches aim to encourage consistency in inference with different levels of perturbations. Our hybrid network achieves desired perturbations not only at the parameter level but also at the architecture level. For a fair comparison, we introduce the UNet [20] as the CNN-based segmentation network and directly explore two successive Swin-ViT layers as a network block to the U-shape segmentation network resulting in a pure ViT-based modified UNet named as SwinUNet [3].

2.3 Scribble-Supervised Loss

To address the challenge of limited-signal scribble-based supervision, CrossEntropy *CE* function is applied solely on the annotated pixels, while ignoring unlabeled pixels as partial supervised segmentation loss (seen in Eq. 2). In this way, we introduce Partial Cross-Entropy *pCE* while only scibble signal training networks [13].

$$\mathcal{L}_{\text{pCE}}(y_{pred}, y_{scrib}) = -\sum_{i\in\omega_L}\sum_{k} y_{scrib}[i,k]log(y_{pred}[i,k]) \quad (2)$$

where i indicates the i-th pixel, and ω_L refers to the set of labeled pixels with scribble annotations. k indicates the k-th class, and $[i, k]$ indicates the probability of i-th pixel belongs to the k-th class.

2.4 Internal Self-Ensembling Consistency Supervision

To boost the performance of each network, we introduce Mean Teacher [22] method from limited-signal SSL task to the similar WSL scribble-supervision task as internal Self-Ensembling consistency supervision (denoted by --→ in Fig. 2). An additional network $f(\overline{\theta})$ with the same architecture but updated by the other network $f(\theta)$ through Exponential Moving Average (EMA) is utilized (as shown in Eq. 3).

$$\overline{\theta_i} = \alpha\theta_i + (1 - \alpha)\theta_{i-1} \tag{3}$$

where θ is the parameter set of the segmentation network at training step i, and $\alpha \in [0, 1]$ is used to balance the weight of updating. Following the internal-consistency requirement, the Gaussian perturbation is applied during training; thus the inference by the Student is enforced to be similar to the Teacher from the same input with noise via internal-consistency loss L_{inter} (as shown in Eq. 4):

$$\mathcal{L}_{\text{inter}}(y_s, y_t) = CE(y_s, y_t) + Dice(y_s, y_t) \tag{4}$$

where *CE* and *Dice* indicate Cross-Entropy-based and Dice-Coefficient-based segmentation loss on the dense pseudo label provided by the Teacher.

2.5 External Dynamic Cross-Consistency Supervision

To ensure that the multiple networks benefit each other, we propose external dynamic cross-consistency supervision. Inspired by MixUp [18,30], we densely combine (in the sense of DenseNet [10]) the output of our group of networks as a dense pseudo-label (in the sense of full segmentation mask), in order to supervise each network iteratively. The dense pseudo annotation, which provides a full-signal supervision, is formulated as:

$$y_{pseudo} = argmax[\frac{1}{2}(\beta y^t_{\text{cnn}} + (1 - \beta)y^s_{cnn}) + \frac{1}{2}(\beta y^t_{\text{vit}} + (1 - \beta)y^s_{\text{vit}}] \tag{5}$$

where y refers to the inference by CNN- and ViT-based networks $f_{vit}(\overline{\theta})$, $f_{vit}(\theta)$, $f_{cnn}(\overline{\theta})$, $f_{cnn}(\theta)$ to provide dense pseudo-labels through dense combinations of output from each model. $\beta \in [0, 1]$ is randomly generated and considered as a kind of 'dynamic' enhanced data perturbation. This process is iterative, thus y_{pseudo} is utilized for network training per iteration as external-consistency loss L_{exter} (seen in Eq. 6):

$$\mathcal{L}_{\text{exter}}(y_{pseudo}, y_{pred}) = CE(y_{pseudo}, y_{pred}) + Dice(y_{pseudo}, y_{pred}) \tag{6}$$

where $CE, Dice$ indicates as Cross-Entropy and Dice-Coefficient-based segmentation loss on the dense pseudo label provided by the dynamic pseudo label ensembling (denoted as --→ in Fig. 2).

3 Experiments

Data Set. We evaluate our proposed CHNets against other baseline methods as a 2D semantic segmentation task on a public benchmarking dataset ACDC [2], a Cardiac MRI segmentation set of 100 patients within five groups: normal controls, heart failure with infarction, dilated cardiomyopathy, hypertrophic cardiomyopathy, abnormal right ventricle. Dense annotations are for Right Ventricle (RV), Myocardium (Myo), and Left Ventricle (LV). Following prior scribble annotation work [1,23], in the pre-processing step we generate the scribble based on the dense annotation already available; we then directly utilize just pure scribbles (and no other annotations) with raw images for training and validation [18]. Dense annotations for the test set enables a conventional validation process to take place by comparing full masks. The data set is randomly splited as 60%, 20% and 20% for training, validation, and testing with no overlap and only conducted once for all baseline and CHNets methods. All images are resized to 224×224 to align with the ViT input style.

Implementation Details. The original UNet [20], and SwinUNet [3] are utilized as CNN- and ViT-based segmentation backbone for CHNets and all WSL baseline methods. The code is written in Pytorch on a single NVIDIA GeForce RTX 3090 GPU, and Intel Core i9-10900K CPU. The dataset is preprocessed for a 2D segmentation task. The total parameter count of ViT- and CNN-based networks is 27.15×10^6 and $1{,}81 \times 10^6$ respectively. The training has 60,000 iterations, the batch size is 12, and the optimizer is SGD (learning rate = 0.1, momentum = 0.9, weight decay = 0.0001). The memory cost is 7 GB, and the runtimes averaged around 4.5 h. The network which performed best on the validation set is utilized for the final testing.

Evaluation Metrics. Various evaluation metrics are utilized to validate the CHNets against other baseline methods including Dice, IOU, Accuracy (Acc), Precision (Pre), Sensitivity (Sen), Specificity (Spe), Hausdorff Distance (HD) in *mm* with a 95% threshold, Average Surface Distance (ASD) in *mm*. Each

Table 1. Direct Comparison of Weakly-Supervised Frameworks on the Test Set

WSL	Net	mDice↑	mIOU↑	Acc↑	Pre↑	Sen↑	Spe↑	HD↓	ASD↓
[13]	ViT	0.8459	0.7355	0.9954	0.8324	0.8709	0.9975	28.6010	7.3933
[14]	ViT	0.8745	0.7802	0.9959	0.8648	0.8920	0.9977	13.4157	3.6616
[12]	ViT	0.8641	0.7630	0.9960	0.8704	0.8655	**0.9982**	6.4881	1.7645
[11]	ViT	0.8632	0.7614	0.9960	**0.8718**	0.8620	**0.9982**	7.6870	2.2027
[21]	ViT	0.8493	0.7405	0.9955	0.8475	0.8678	0.9978	8.3234	2.3858
[13]	CNN	0.6455	0.4918	0.9831	0.5318	0.8945	0.9848	163.5975	69.0296
[14]	CNN	0.8588	0.6147	0.9904	0.6501	**0.9203**	0.9916	143.5347	44.8333
[12]	CNN	0.8645	0.7644	0.9955	0.8449	0.8904	0.9973	28.4650	7.6293
[11]	CNN	0.8681	0.7709	0.9957	0.8518	0.8915	0.9975	23.6676	6.6040
[21]	CNN	0.8709	0.7755	0.9957	0.8519	0.9030	0.9974	7.8396	1.8412
Ours		**0.8906**	**0.8058**	**0.9964**	0.8698	<u>0.9158</u>	<u>0.9978</u>	**5.4180**	**1.6484**

Table 2. Direct Comparison of Weakly-Supervised Frameworks on the Test Set of Each Segmented Feature

WSL	Net	RV			Myo			LV		
		Dice↑	HD↓	ASD↓	Dice↑	HD↓	ASD↓	Dice↑	HD↓	ASD↓
[13]	ViT	0.8587	9.3925	3.7748	0.7859	45.0363	10.0612	0.8929	31.3743	8.3438
[14]	ViT	0.8639	9.4354	2.9105	0.8230	14.83338	4.3353	0.9366	15.9782	3.7390
[12]	ViT	0.8727	6.7018	1.6205	0.8105	5.7516	1.5848	0.9091	7.0109	2.0971
[11]	ViT	0.8678	**6.9280**	**1.6073**	0.8137	7.1041	2.1839	0.9081	9.0291	2.8168
[21]	ViT	0.8622	6.9086	1.7250	0.7904	8.0107	2.2518	0.8952	10.0510	3.1806
[13]	CNN	0.5806	182.2923	87.5389	0.5260	160.3049	68.6412	0.8300	163.5975	69.0296
[14]	CNN	0.7304	138.4518	41.0612	0.7102	125.1634	31.8241	0.8360	166.9888	61.6147
[12]	CNN	0.8502	11.2341	3.3072	0.8156	29.3005	8.0682	0.9276	44.8603	11.5125
[11]	CNN	0.8354	29.2791	8.0856	0.8260	24.3843	7.6606	0.9427	17.3394	4.0656
[21]	CNN	0.8519	13.5882	3.1754	0.8164	3.8603	**1.2166**	0.9444	6.0701	1.1317
Ours		**0.8752**	8.9538	2.3428	**0.8445**	**3.6503**	<u>1.5336</u>	**0.9519**	**3.6499**	**1.0687**

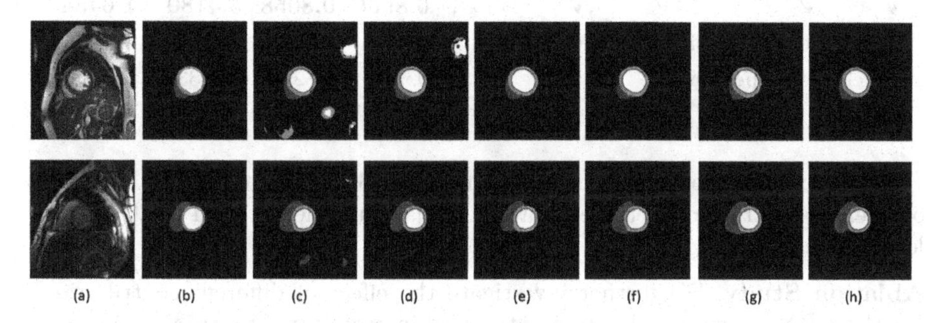

Fig. 3. Two Sample ACDC Images with Their Corresponding Inferences. Left to right, (a) Input MRI raw image, (b) Ground Truth. Further, the images show inferences from (c) pCE [13], (d) USTM [14], (e) Scribble2Label [12], (f) Mumford-Shah loss [11], (g) GatedCRFLoss [21] and (h) CHNets (ours).

metric is annotated with ↑ or ↓ to indicate whether higher is better or lower is better. mDice in Table 1 and Dice in Table 2 respectively refer to the mean Dice-coefficient over the three classes (RV, Myo and LV), or per class.

Comparison with Baseline Methods. CHNets is compared against five other scribble-supervised segmentation methods including pCE [13], USTM [14], Scribble2Label [12], Mumford-Shah loss [11], and GatedCRFLoss [21]. All baseline methods and CHNets are trained with the same hyper-parameter setting, the same loss functions (*pCE*), and the same quality of scribble annotations. Each of the scribble-supervised methods is extended to be with either CNN- or ViT-based network as the segmentation backbone. The quantitative results of the direct comparison of baseline methods and CHNets are reported in Table 1, and we further report the performance of each region of interest in Table 2. The best performance is highlighted with **Bold**, and the second best is <u>Underlined</u>.

Table 3. Ablation Study with Two Internal-External Supervision on the Test Set and Fully Supervised Supervision

Consistency-Aware Supervision				Performance			
Internal Consistency		External Consistency		mDice↑	mIOU↑	HD↓	ASD↓
CNN	ViT	CNN	ViT				
✓ × 2				0.7083	0.5544	150.5851	50.3175
	✓ × 2			0.7612	0.6253	148.5577	43.7664
		✓ × 2		0.8837	0.7945	6.1310	4.9041
			✓ × 2	0.7392	0.6087	62.4700	24.7017
✓ × 2		✓		0.8846	0.7964	8.2995	2.8425
	✓ × 2	✓		0.8880	0.8012	12.2475	3.2928
✓ × 2			✓	0.8815	0.7902	12.7286	3.7176
	✓ × 2		✓	0.8633	0.7632	7.3206	2.4864
✓	✓	✓	✓	**0.8906**	**0.8058**	**5.4180**	**1.6484**
Pixel-level Supervision			CNN	0.9167	0.9120	3.7452	0.8615
Pixel-level Supervision			ViT	0.9049	0.8290	3.6233	0.8749

The qualitative performance is shown in Fig. 3 where the outside boundary of inference on the test set is evaluated against published ground truth at pixel level of 4 classes.

Ablation Study. We further investigate the effect of different contributions for the CHNets, where all combinations of internal-consistency and external-consistency, with CNN or ViT chosen as backbone, are reported in Table 3. The symbol ✓ × 2 indicates two models for the internal-consistency training with self-ensembling fashion. A tick ✓ with all internal-consistency and external-consistency with CNN and ViT refers to CHNets which achieve the best performance demonstrating the promising improvement by our proposed techniques. The pixel-level ground truth supervised training with CNN and ViT is further reported as the upper bound of performance, and we find our proposed method has achieved very similar results compared with the upper-bound performance.

Clinical Application. It is important to emphasise that the numerical evaluation measures are only indicative of the power of such methods. To identify cardiomyopathy, for instance, clinicians are interested in the strength of the Myocardium. This, in turn, is indicated by the circularity of the Left Ventricle. A precise segmentation of the LV is, therefore, not essential: what is important is to gauge its correct xy aspect ratio, which can be obtained by scribble-training.

4 Conclusion

The CNN and ViT architectures have been developed and trained simultaneously in an end-to-end manner. Internal- and external-consistency training schemes are proposed to boost the performance of each network and benefit each other. The quantitative experiments on the public benchmark MRI dataset demonstrate promising performance of the proposed method against other scribble-supervised methods. In future, we will extend experiments to other limited-signal supervision for training such as bounding boxes or point-based annotations.

References

1. Bai, W., et al.: Semi-supervised learning for network-based cardiac MR image segmentation. In: Descoteaux, M., Maier-Hein, L., Franz, A., Jannin, P., Collins, D.L., Duchesne, S. (eds.) MICCAI 2017. LNCS, vol. 10434, pp. 253–260. Springer, Cham (2017). https://doi.org/10.1007/978-3-319-66185-8_29
2. Bernard, O., et al.: Deep learning techniques for automatic MRI cardiac multi-structures segmentation and diagnosis: is the problem solved? IEEE TMI (2018)
3. Cao, H., et al.: Swin-unet: unet-like pure transformer for medical image segmentation. arXiv preprint arXiv:2105.05537 (2021)
4. Chen, L.C., et al.: Deeplab: semantic image segmentation with deep convolutional nets, atrous convolution, and fully connected CRFs. IEEE TPAMI (2017)
5. Chen, X., Yuan, Y., Zeng, G., Wang, J.: Semi-supervised semantic segmentation with cross pseudo supervision. In: Proceedings of the IEEE/CVF Conference on Computer Vision and Pattern Recognition, pp. 2613–2622 (2021)
6. Dosovitskiy, A., et al.: An image is worth 16 × 16 words: transformers for image recognition at scale. arXiv preprint arXiv:2010.11929 (2020)
7. Grady, L.: Random walks for image segmentation. IEEE Trans. Pattern Anal. Mach. Intell. (2006)
8. Grandvalet, Y., Bengio, Y.: Semi-supervised learning by entropy minimization. NIPS (2004)
9. Han, B., et al.: Co-teaching: robust training of deep neural networks with extremely noisy labels. NIPS (2018)
10. Huang, G., Liu, Z., Van Der Maaten, L., Weinberger, K.Q.: Densely connected convolutional networks. In: Proceedings of the IEEE Conference on Computer Vision and Pattern Recognition (CVPR), pp. 4700–4708 (2017)
11. Kim, B., Ye, J.C.: Mumford-shah loss functional for image segmentation with deep learning. IEEE TIP (2019)
12. Lee, H., Jeong, W.-K.: Scribble2Label: scribble-supervised cell segmentation via self-generating pseudo-labels with consistency. In: Martel, A.L., et al. (eds.) MICCAI 2020. LNCS, vol. 12261, pp. 14–23. Springer, Cham (2020). https://doi.org/10.1007/978-3-030-59710-8_2
13. Lin, D., Dai, J., Jia, J., He, K., Sun, J.: Scribblesup: scribble-supervised convolutional networks for semantic segmentation. In: Proceedings of the IEEE Conference on Computer Vision and Pattern Recognition, pp. 3159–3167 (2016)
14. Liu, X., et al.: Weakly supervised segmentation of COVID19 infection with scribble annotation on CT images. Pattern Recognit. (2022)
15. Liu, Z., et al.: Swin transformer v2: scaling up capacity and resolution. In: Proceedings of the IEEE/CVF Conference on Computer Vision and Pattern Recognition, pp. 12009–12019 (2022)
16. Liu, Z., et al.: Swin transformer: hierarchical vision transformer using shifted windows. In: Proceedings of the IEEE/CVF International Conference on Computer Vision, pp. 10012–10022 (2021)

17. Luo, X., Hu, M., Song, T., Wang, G., Zhang, S.: Semi-supervised medical image segmentation via cross teaching between CNN and transformer. In: International Conference on Medical Imaging with Deep Learning, pp. 820–833. PMLR (2022)
18. Luo, X., et al.: Scribble-supervised medical image segmentation via dual-branch network and dynamically mixed pseudo labels supervision. In: Wang, L., Dou, Q., Fletcher, P.T., Speidel, S., Li, S. (eds.) Medical Image Computing and Computer Assisted Intervention – MICCAI 2022. MICCAI 2022. LNCS, vol. 13431, pp. 528–538. Springer, Cham (2022). https://doi.org/10.1007/978-3-031-16431-6_50
19. Reiß, S., Seibold, C., Freytag, A., Rodner, E., Stiefelhagen, R.: Every annotation counts: multi-label deep supervision for medical image segmentation. In: Proceedings of the IEEE/CVF Conference on Computer Vision and Pattern Recognition, pp. 9532–9542 (2021)
20. Ronneberger, O., Fischer, P., Brox, T.: U-Net: convolutional networks for biomedical image segmentation. In: Navab, N., Hornegger, J., Wells, W.M., Frangi, A.F. (eds.) MICCAI 2015. LNCS, vol. 9351, pp. 234–241. Springer, Cham (2015). https://doi.org/10.1007/978-3-319-24574-4_28
21. Tang, M., Perazzi, F., Djelouah, A., Ayed, I.B., Schroers, C., Boykov, Y.: On regularized losses for weakly-supervised CNN segmentation. In: Ferrari, V., Hebert, M., Sminchisescu, C., Weiss, Y. (eds.) ECCV 2018. LNCS, vol. 11220, pp. 524–540. Springer, Cham (2018). https://doi.org/10.1007/978-3-030-01270-0_31
22. Tarvainen, A., Valpola, H.: Mean teachers are better role models: weight-averaged consistency targets improve semi-supervised deep learning results. Adv. Neural Inf. Process. Syst. **30** (2017)
23. Valvano, G., et al.: Learning to segment from scribbles using multi-scale adversarial attention gates. IEEE TMI (2021)
24. Vaswani, A., et al.: Attention is all you need. NIPS (2017)
25. Wang, Z., et al.: Uncertainty-aware transformer for MRI cardiac segmentation via mean teachers. MIUA (2022)
26. Wang, Z., Dong, N., Voiculescu, I.: Computationally-efficient vision transformer for medical image semantic segmentation via dual pseudo-label supervision. In: ICIP. IEEE (2022)
27. Wang, Z., Voiculescu, I.: Triple-view feature learning for medical image segmentation. In: Xu, X., Li, X., Mahapatra, D., Cheng, L., Petitjean, C., Fu, H. (eds.) Resource-Efficient Medical Image Analysis. REMIA 2022. LNCS, vol. 13543, pp. 42–54. Springer, Cham (2022). https://doi.org/10.1007/978-3-031-16876-5_5
28. Wang, Z., Voiculescu, I.: Dealing with unreliable annotations: a noise-robust network for semantic segmentation through a transformer-improved encoder and convolution decoder. Appl. Sci. (2023)
29. Wang, Z., Zhang, Z., Voiculescu, I.: RAR-U-Net: a residual encoder to attention decoder by residual connections framework for spine segmentation under noisy labels. In: ICIP. IEEE (2021)
30. Zhang, H., et al.: Mixup: beyond empirical risk minimization. ICLR (2017)
31. Zhang, K., Zhuang, X.: Cyclemix: a holistic strategy for medical image segmentation from scribble supervision. In: Proceedings of the IEEE/CVF Conference on Computer Vision and Pattern Recognition, pp. 11656–11665 (2022)

Whole Slide Multiple Instance Learning for Predicting Axillary Lymph Node Metastasis

Glejdis Shkëmbi[1(✉)], Johanna P. Müller[1], Zhe Li[1], Katharina Breininger[1], Peter Schüffler[2], and Bernhard Kainz[1,3]

[1] Friedrich-Alexander-Universität Erlangen-Nürnberg, Erlangen, Germany
glejdisshkembi@gmail.com , johanna.paula.mueller@fau.de
[2] Technical University of Munich, Munich, Germany
[3] Imperial College London, London, UK

Abstract. Breast cancer is a major concern for women's health globally, with axillary lymph node (ALN) metastasis identification being critical for prognosis evaluation and treatment guidance. This paper presents a deep learning (DL) classification pipeline for quantifying clinical information from digital core-needle biopsy (CNB) images, with one step less than existing methods. A publicly available dataset of 1058 patients was used to evaluate the performance of different baseline state-of-the-art (SOTA) DL models in classifying ALN metastatic status based on CNB images. An extensive ablation study of various data augmentation techniques was also conducted. Finally, the manual tumor segmentation and annotation step performed by the pathologists was assessed. Our proposed training scheme outperformed SOTA by 3.73%. Source code is available here.

1 Introduction

Breast cancer (BC) is currently the world's most diagnosed cancer [3], accounting for 685,000 deaths among women in 2020. Axillary lymph nodes (ALNs) are typically the first location of breast cancer metastasis, which makes their status the single most important predictive indicator for diagnosis [7]. Several studies have demonstrated that deep learning (DL) can support pathologists by increasing the sensitivity of ALN micro-metastasis detection, as shown in the review paper [10]. Furthermore, earlier research has demonstrated lymph node metastasis detection may be assisted by deep features from whole slide images (WSIs) [13]. Deep learning has greatly reduced the need for domain experts to manually extract features from data. However, human experts are still essential for data labeling. Unfortunately, the increasing complexity of many problems requires large amounts of annotated data, which can be costly and raise privacy concerns since the process of labeling often involves analyzing and categorizing personal information or sensitive data, especially, in a medical context. Pixel-level annotation is a time-consuming and expensive process, as it requires a

B. Bhattarai et al. (Eds.): DEMI 2023, LNCS 14314, pp. 11–20, 2023.
https://doi.org/10.1007/978-3-031-44992-5_2

human annotator to go through every single pixel in an image to label it correctly. This process can take a lot of time and effort, especially for large datasets or complex images, and may also require a high level of expertise or specialized training. On the other hand, image-level annotation is a much simpler and faster process, as it only requires a human annotator to assign a single label to an entire image.

Multiple Instance Learning (MIL) has enabled pathologists to label bags of sub-images or patches, rather than labeling each individual patch. This approach is particularly useful in binary classification tasks, where distinguishing between healthy and diseased patients is the primary goal. By using MIL, an image can be labeled as malignant if it contains at least one malignant patch, while an image is considered cancer-free if all patches are classified as healthy. We present a pipeline that enables the prediction of axillary lymph node metastasis status from whole slide images of core-needle biopsy samples from patients with early breast cancer. The prediction of metastasis status based solely on histopathological images is a complex and challenging task and one that previously exceeded the competencies of medical experts. Our approach surpasses the limitations of traditional manual assessments and pixel-wise annotation of whole slide images, but yet provides a reliable and objective method for predicting metastasis status.

In this paper, we reproduce the results of the attention-based MIL classification model proposed by [12], which aims at identifying the (micro-)metastatic status of ALN preoperatively using the Early Breast Cancer Core-Needle Biopsy WSI (BCNB) [12] dataset of early breast cancer (EBC) patients. To evaluate the impact of feature extraction components on the classification pipeline's performance, we utilized 43 different convolutional neural networks (CNNs) as a backbone for feature extraction. Given the constraints of limited data, we hypothesize that data augmentation can enhance and diversify the original BCNB dataset [12], leading to improved model generalization and overall performance. An extensive ablation study on different data augmentation techniques, including basic and advanced methods, examined this effect. Finally, the pipeline proposed by [12] relied on hand-labeled, pixel-wise annotation of the tumor. However, manual tumor segmentation and annotation is a time-consuming process and prone to errors. As such, we tested the necessity of this information input for the deep-learning core-needle biopsy model.

Related Work: Recently, DL has demonstrated its ability to extract features from medical images at a high throughput rate and analyze the correlation between primary tumor features and ALN metastasis information. In the study by [15], researchers utilized Inception-v3 [11], Inception-ResNet-v2 and ResNet [4] to predict clinically negative ALN metastasis using two-dimensional greyscale ultrasound images of patients with primary breast cancer. Moreover, [14] employed two-dimensional shear wave elastography (SWE), a new ultrasound method for measuring tissue stiffness, to discriminate between malignant and benign breast tumors. They combined clinical parameters with DL radiomics, where a pre-trained ResNet model [4] encoded the input images into features. This combination achieved an area under the receiver operating characteristic

(AUROC) value of 0.902 in the test cohort. Comparatively, their classification model using only clinicopathologic data achieved an AUROC value of 0.727.

2 Method

Our goal in this paper is to predict the status of axillary lymph nodes (ALN) using core needle biopsies. We present the deep learning (DL)-based core-needle biopsy with whole slide processing (DLCNBC-WS) network, a revised method that builds on two attention-based deep MIL frameworks, proposed by [12]. The Deep-learning core-needle biopsy (DLCNB) model is the base model for both frameworks and is built on an attention-based deep MIL framework for predicting the ALN status using DL (histopathological) features [6], from core-needle biopsies. These DL features were extracted only from the cancerous areas of the WSIs of breast CNB samples. We differentiate between non-existing or already advanced metastasis as ALN status. In contrast to the DLCNB model, the DLCNBC model uses additionally selected features from the clinical data set. The clinical information of the slide is added to all the constructed bags in order to provide insightful information for predicting and achieving better performance. The whole algorithm pipeline is composed of four steps and follows [12].

One of the key features of our revised method is the eliminated segmentation of the cancerous features in the CNB slides which allows higher flexibility and adaptability. Our proposed DLCNBC with whole slide processing (DLCNBC-WS) omits the time-consuming step of segmentation and reduces the requirements of its applications. We accomplish to cut the steps of the algorithm pipeline into only three remaining steps.

In the following, we explain the DLCNBC-WS in detail, see Fig. 1. First, N feature vectors for the N WSI patches of size 256×256 pixels in each bag were extracted using multiple different CNN models as the backbone for feature extraction. Multiple bags were built for each WSI (top image of Fig. 1). As the bottom image of Fig. 1 shows, the clinical data were also preprocessed for feature extraction in addition to the WSIs. For tackling the restricted availability of the training data and therefore the issue of overfitting, a wide range of basic and advanced data augmentation techniques were applied to the WSI patches.

Next, the N feature vectors of patches in a bag were processed by max-pooling and reshaping. Then, they were passed through two fully connected layers to output N weight factors for patches in the bag, which were then further multiplied by the original feature vectors in order to dynamically adjust the impact of instance features. Finally, the weighted image feature vectors, created by the aggregation of features of multiple patches from the attention module, were fused with clinical features by concatenation. A classifier learns to estimate ALN status based on aggregated input.

3 Evaluation

Data Source: The BCNB dataset used in this paper includes clinical data and the core-needle biopsy (CNB) hematoxylin and eosin (H&E) stained WSIs of

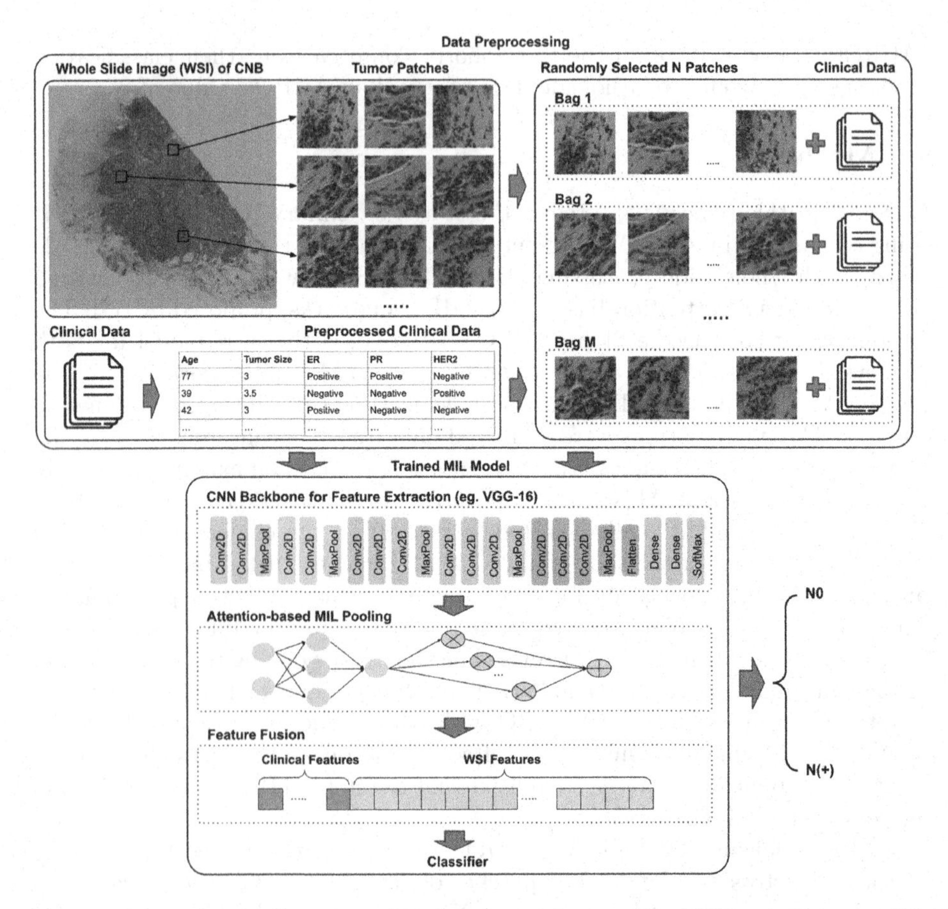

Fig. 1. The overall pipeline of DLCNBC-WS model to predict ALN status between N0 and N(+). Top image: Multiple training bags were built. Bottom image: DLCNBC-WS model training process included feature extraction and MIL, and by aggregating the model outputs of all bags from the same slide, the ALN state is predicted.

patients with pathologically confirmed ALN status. It includes the following clinical features: age, tumor size, tumor type, estrogen receptor (ER)-status, progesterone receptor (PR)-status, human epidermal growth factor receptor 2 (HER2)-status, HER2 expression, histological grading, surgical procedure (ALND or SLNB), Ki67, molecular subtype, the number of lymph node metastases (LNM) and the target variable which is the metastatic status of ALN (N0, N+(1-2) or N+(>2)). The dataset contains 655 patients with N0 status (no metastasis), 210 patients with N+(1-2) status (cancer has metastasized to 1–2 ALNs), and 193 patients with N+(>2) ALN status (cancer has metastasized to 3 or more ALNs).

Preprocessing: The selected numerical features (age and tumor size) of the clinical dataset were standardized and the selected categorical variables (ER/ PR/ HER-2 status and the ALN metastases status as label) were one-hot encoded as no ordinal relationship was observed. The WSI preparation was performed according to [12]. The multiple training bags were constructed by fusing together the features extracted from the cropped patches of the selected tumor regions of each CNB WSI with the selected features from the clinical dataset. The WSIs and the clinical data were divided on a per-patient level into training and independent testing cohorts; 80:20 respectively, according to [12], where one WSI belongs to one patient and clinical information of each patient is represented by one entry in the clinical dataset. The validation cohort was created from the training set by randomly selecting 25% of that data. In this way, the patches from one patient are found exclusively in either the train/validation set or the test set, but not in both. In [12], the model relied on manually annotated cancerous regions in each WSI. However, to reduce the need for expert pathologists, decrease human errors in annotation, and increase reproducibility, patches were sampled from all (tissue) areas of the WSI, in contrast to the pipeline by [12]. They extracted the patches from only tumor areas which were manually selected in advance by pathologists. Our pipeline framework assumes that the model has no prior knowledge of whether the sampled patch belongs to a tumor. For adapting the dataset to our proposed pipeline, the sampled patches with artefacts or without cell information were filtered out using Shannon entropy computed on the grey-scale versions of the patches.

Training: The model is trained to predict the ALN metastasis based on WSIs of primary BC samples by predicting the bag label while thoroughly considering all included instances in each bag. A stochastic gradient descent (SGD) optimizer with a learning rate of $1e^{-4}$ to update the model parameters and a cosine annealing warm restarts strategy to adjust the learning rate were utilized. In the training phase, L2 regularization was implemented by passing weight decay of $1e^{-3}$ to the optimizer. In addition, to ease overfitting L1 regularization was employed with weight $1e^{-6}$. During the testing phase, the ALN status was predicted by aggregating the model outputs of all bags from the same WSI (Fig. 1c). We trained our models on NVIDIA RTX A6000 for 2000 epochs.

Statistical Analysis: To examine the clinical characteristics of different cohorts, we conducted a correlation analysis. To establish a basis for comparison, we first utilized all features of the clinical dataset to develop a logistic regression (LR) model for ALN prediction. We then trained the LR model using a subset of features from the clinical dataset, specifically age, tumor size, and the expression levels of ER, PR, and HER2 biomarkers. These specific features were chosen based on the results of a correlation analysis conducted on the clinical dataset, which involved selecting features with a higher correlation to the outcome variable and a lower correlation among other predicting variables. The reduced clinical dataset was then incorporated into the input of the DLCNBC model.

Ablation Study: Comparing multiple backbone models is essential to building a robust and effective deep-learning classification pipeline. We compared multiple CNN models for finding the best-suited pipeline for this specific task. We updated the list of backbone models with the latest advances: AlexNet, VGG with and without BN, GoogleNet, Inception-v3, ResNet, DenseNet, SqueezeNet, ResNext, WideResNet, MobileNet, ShuffleNet-v2, EfficientNet, and EfficientNet-v2.

For identifying which augmentation techniques contribute the most to improving the model's accuracy and generalization ability, we conducted an extensive ablation study. All augmentations techniques were applied as a preprocessing step on the training cohort of the WSI patches. For model training, each image was resized to 224×224 pixels and normalized. Next, different data augmentation methods, including different types of geometric augmentation (including flipping, rotation, translation, shearing and scaling), colors space augmentation (including changes in brightness, contrast, saturation and hue, converting an image to grayscale, solarization, and posterization), image erasing (including random erasing and random cropping), AugMix [5], as well as more advanced approaches such as AutoAugment [1], RandAugment [2] and TrivialAugmentWide [9], were applied. Basic augmentation techniques such as geometric augmentation, color space augmentation, image erasing and image mixing were manually designed, and a manually predefined set of parameter values was applied. The augmentation techniques that yielded the best performance on the test dataset, as determined by AUROC, were selected and combined into an augmentation mixture, which was subsequently evaluated to assess its impact on model performance. The best-performing model based on [12] (DLCNBC model with VGG-16 BN) was trained on each data augmentation technique. Finally, the DLCNBC and DLCNBC-WS models were trained, as described above.

4 Results and Discussion

In this section, we present and analyze the results of our experiments. First, the performance results in terms of AUROC, accuracy, sensitivity, specificity, PPV, NPV, and F1-score in the test cohort of the best and worst-performing DLCNBC models with different backbones for feature extraction in the binary classification of ALN status are illustrated in Table 1. The highest AUROC score of 0.837 was achieved using VGG-13 with batch normalization (BN) as the feature extractor, while the lowest AUROC score of 0.543 was achieved by EfficientNet-b0, EfficientNet-b2 and EfficientNet-b3. In general, EfficientNet and Inception-v3 have shown not to outperform VGG in terms of accuracy and AUROC while using fewer computational resources. However, VGG outperformed all other backbone models, for which its reasons needs to be further investigated in future. Second, the ablation study showed that the classification was improved by further applying one of the following augmentation strategies to the data for DLCNBC models: random rotation, scaling, shearing, or random vertical flipping, see Table 2. The best results were obtained by performing a random rotation $\leq 10°$. It increased the AUROC score from 0.837 to 0.852 in the binary classification task.

Third, our proposed DLCNBC-WS model performed best when VGG-13 BN was used as the backbone for feature extraction in the binary classification task

Table 1. The performance results on the test cohort of the best and worst performing DLCNBC models with different backbones for feature extraction in the binary classification of ALN status (N0 vs. N(+)).

Backbone	AUROC	Accuracy	Sensitivity	Specificity	PPV	NPV	F1-Score
VGG-11 BN	0.83	**0.739**	0.607	0.821	0.68	0.769	0.642
VGG-13 BN	**0.837**	**0.739**	0.548	0.858	0.708	0.752	0.617
VGG-16 BN	0.822	**0.739**	0.679	0.776	0.655	0.794	0.667
VGG-19 BN	0.828	**0.739**	**0.881**	0.649	0.612	**0.897**	**0.722**
Inception-v3	0.545	0.633	0.095	0.970	0.667	0.631	0.167
EfficientNet-b0	0.543	0.624	0.024	**1.000**	**1.000**	0.620	0.047
EfficientNet-b2	0.545	0.615	0.048	0.970	0.500	0.619	0.087
EfficientNet-b3	0.543	0.615	0.071	0.955	0.500	0.621	0.125

Table 2. The performance results on the test cohort of the four best and worst performing data augmentation techniques on the DLCNBC models with VGG-16 BN as the backbone for feature extraction in the binary classification of ALN status (N0 vs. N(+)).

Augmentation	AUROC	Accuracy	Sensitivity	Specificity	PPV	NPV	F1-Score
Random Rotation	**0.868**	0.761	0.690	0.806	0.690	0.806	0.690
Shear	0.865	0.775	0.679	**0.836**	**0.722**	0.806	0.699
Random Erasing	0.863	0.757	**0.821**	0.716	0.645	**0.865**	0.723
Vertical Flip	0.857	**0.780**	0.762	0.791	0.696	0.841	**0.727**

Table 3. Comparison of the performance of DLCNBC and DLCNBC-WS models with VGG-13 BN backbone for binary classification of the ALN status. T: training cohort, V: validation cohort, I-T: independent test cohort.

Method	Set	AUROC	Accuracy	Sensitivity	Specificity	PPV	NPV	F1-Score
DLCNBC	T	0.903	0.814	0.812	0.815	0.73	0.876	0.769
	V	0.847	0.767	0.734	0.786	0.674	0.831	0.703
	I-T	0.837	0.739	0.548	**0.858**	0.708	0.752	0.617
DLCNBC Rotation	T	0.945	0.881	0.896	0.872	0.811	0.932	0.851
	V	0.866	0.790	0.886	0.733	0.667	0.914	0.761
	I-T	0.852	0.766	0.774	0.761	0.670	0.843	0.718
DLCNBC-WS	T	0.983	0.938	0.941	0.935	0.899	0.963	0.920
	V	0.863	0.781	0.823	0.756	0.670	0.876	0.739
	I-T	**0.862**	**0.803**	**0.833**	0.784	0.707	**0.882**	**0.765**
DLCNBC-WS Rotation	T	0.976	0.910	0.907	0.912	0.863	0.941	0.885
	V	0.884	0.790	0.709	0.840	0.727	0.827	0.718
	I-T	0.843	0.766	0.619	**0.858**	**0.732**	0.782	0.671

N0 vs. N(+), see Table 4, the results for multi-class classification are given in Table 5.

Table 3 compares the results of the baseline method DLCNBC and our proposed DLCNBC-WS with the best-performing backbone model for binary clas-

Table 4. The performance results on the test cohort of the DLCNBC-WS models with different backbones for feature extraction in the binary classification of ALN status (N0 vs. N(+)).

Backbone	AUROC	Accuracy	Sensitivity	Specificity	PPV	NPV	F1-Score
VGG-11	0.82	0.711	0.655	0.746	0.618	0.775	0.636
VGG-11 BN	**0.86**	0.739	0.940	0.612	0.603	**0.943**	0.735
VGG-13	0.84	0.78	0.690	0.836	0.725	0.812	0.707
VGG-13 BN	**0.86**	**0.803**	**0.833**	0.784	0.707	0.882	**0.765**
VGG-16	0.85	0.766	0.571	0.888	**0.762**	0.768	0.653
VGG-16 BN	**0.86**	0.706	0.357	**0.925**	0.750	0.697	0.484
VGG-19	0.83	0.784	0.738	0.813	0.713	0.832	0.725
VGG-19 BN	0.85	0.766	0.571	0.888	**0.762**	0.768	0.653

Table 5. The performance results for each class on the test cohort of the best and worst performing DLCNBC models with different backbones for feature extraction in the multi-class classification of ALN status (N0 vs. N+(1-2) vs. N+(>2)).

Backbone	Class	AUROC	Accuracy	Sensitivity	Specificity	PPV	NPV
VGG-11 BN	N0	0.84	0.761	0.855	0.621	**0.772**	0.740
	N+(1-2)	0.80	0.784	0.440	0.887	0.537	0.842
	N+(>2)	0.68	0.766	0.243	0.873	0.281	0.849
VGG-13 BN	N0	0.84	0.757	0.847	0.621	0.771	0.730
	N+(1-2)	0.79	0.784	0.420	0.893	0.538	0.838
	N+(>2)	0.71	0.761	0.270	0.862	0.286	**0.852**
VGG-16 BN	N0	**0.85**	0.748	0.931	0.471	0.726	0.820
	N+(1-2)	0.79	0.784	0.360	0.911	0.545	0.827
	N+(>2)	0.71	0.807	0.162	0.939	0.353	0.846
WideResNet-101-2	N0	0.58	0.587	0.939	0.057	0.600	0.385
	N+(1-2)	0.60	0.775	0.060	**0.988**	0.600	0.779
	N+(>2)	0.48	0.794	0.000	0.956	0.000	0.824
MobileNet v3 large	N0	0.64	0.615	**0.947**	0.115	0.617	0.588
	N+(1-2)	0.60	0.775	0.120	0.970	0.545	0.787
	N+(>2)	0.50	**0.812**	0.027	0.972	0.167	0.830
EfficientNet b3	N0	0.63	0.606	**0.947**	0.092	0.611	0.533
	N+(1-2)	0.59	0.780	0.120	0.976	0.600	0.788
	N+(>2)	0.51	0.807	0.000	0.972	0.000	0.826
EfficientNet b5	N0	0.64	0.601	0.916	0.126	0.612	0.500
	N+(1-2)	0.62	0.752	0.100	0.946	0.357	0.779
	N+(>2)	0.48	**0.812**	0.054	0.967	0.250	0.833

sification, VGG-13 with BN. We also show the impact of random rotation as an augmentation function for both models. The table shows that applying random rotation increased the model performance in terms of AUROC score only for the baseline method by 1.79%. Moreover, training the model without the expert tumor segmentation step surpassed the results of the preceding paper on the BCNB dataset [12] for the independent test cohort by 3.37% in the AUROC score.

5 Conclusion

In this paper, we succeeded in reproducing the results of [12] and further upgraded their attention-based MIL classification pipeline for predicting ALN metastasis status preoperatively in EBC patients. We found that both the performance of DLCNB and DLCNBC were significantly influenced by CNN backbone selection. The results of the ablation study showed that learning was heavily influenced by the preparation of training data and changes in data distribution. In particular, the use of random rotation on top of the baseline model outperformed SOTA in the binary classification of ALN status by 2.53%. Lastly, the requirement of pathologists to perform tumor segmentation and annotation was removed and SOTA was outperformed by 3.73%.

Since the waiting time between biopsy and pathological classification affects whether a diagnosis of lymph node metastasis still reflects the current status and, hence, is accurate, the ALN metastasis is intrinsically unstable [15]. For instance, if monitored for a sufficient amount of time, some patients with negative lymph nodes may eventually develop positive lymph nodes. In addition, an interesting attempt would be to evaluate the clinical utility of immunochemically stained images, instead of focusing solely on H&E staining. One important limitation of CNNs in histopathology image analysis does not clearly capture inter-nuclear interactions and histopathological information. This is important for detecting and characterizing cancers [8], which could be solved with the help of graph neural networks.

Acknowledgements. The authors gratefully acknowledge the scientific support and HPC resources provided by the Erlangen National High Performance Computing Center (NHR@FAU) of the Friedrich-Alexander-Universität Erlangen-Nürnberg (FAU) under the NHR projects b143dc and b180dc. NHR funding is provided by federal and Bavarian state authorities. NHR@FAU hardware is partially funded by the German Research Foundation (DFG) - 440719683. Additional support was also received by the ERC - project MIA-NORMAL 101083647, DFG KA 5801/2-1, INST 90/1351-1 and by the state of Bavaria.

References

1. Cubuk, E.D., Zoph, B., Mane, D., Vasudevan, V., Le, Q.V.: Autoaugment: learning augmentation policies from data (2018). https://doi.org/10.48550/ARXIV.1805.09501, https://arxiv.org/abs/1805.09501
2. Cubuk, E.D., Zoph, B., Shlens, J., Le, Q.V.: Randaugment: practical automated data augmentation with a reduced search space (2019). https://doi.org/10.48550/ARXIV.1909.13719, https://arxiv.org/abs/1909.13719

3. Ferlay, J., et al.: Cancer statistics for the year 2020: an overview. Int. J. Cancer **149**(4), 778–789 (2021). https://doi.org/10.1002/ijc.33588, https://onlinelibrary.wiley.com/doi/abs/10.1002/ijc.33588
4. He, K., Zhang, X., Ren, S., Sun, J.: Deep residual learning for image recognition (2015). https://doi.org/10.48550/ARXIV.1512.03385, https://arxiv.org/abs/1512.03385
5. Hendrycks, D., Mu, N., Cubuk, E.D., Zoph, B., Gilmer, J., Lakshminarayanan, B.: Augmix: a simple data processing method to improve robustness and uncertainty (2019). https://doi.org/10.48550/ARXIV.1912.02781, https://arxiv.org/abs/1912.02781
6. Ilse, M., Tomczak, J.M., Welling, M.: Attention-based deep multiple instance learning. In: Proceedings of the 35th International Conference on Machine Learning (ICML) abs/1802.04712 (2018). https://arxiv.org/abs/1802.04712
7. Jatoi, I., Hilsenbeck, S.G., Clark, G.M., Osborne, C.K.: Significance of axillary lymph node metastasis in primary breast cancer (1999). https://doi.org/10.1200/JCO.1999.17.8.2334, https://pubmed.ncbi.nlm.nih.gov/10561295/
8. Jaume, G., et al.: Quantifying explainers of graph neural networks in computational pathology. In: IEEE/CVF Conference on Computer Vision and Pattern Recognition (CVPR) abs/2011.12646 (2020). https://arxiv.org/abs/2011.12646
9. Müller, S.G., Hutter, F.: Trivialaugment: tuning-free yet state-of-the-art data augmentation (2021). https://doi.org/10.48550/ARXIV.2103.10158, https://arxiv.org/abs/2103.10158
10. Steiner, D., et al.: Impact of deep learning assistance on the histopathologic review of lymph nodes for metastatic breast cancer. Am. J. Surg. Pathol. **42**, 1 (2018). https://doi.org/10.1097/PAS.0000000000001151
11. Szegedy, C., Vanhoucke, V., Ioffe, S., Shlens, J., Wojna, Z.: Rethinking the inception architecture for computer vision (2015). https://doi.org/10.48550/ARXIV.1512.00567, https://arxiv.org/abs/1512.00567
12. Xu, F., et al.: Predicting axillary lymph node metastasis in early breast cancer using deep learning on primary tumor biopsy slides. Front. Oncol. **11**, 4133 (2021). https://doi.org/10.3389/fonc.2021.759007, https://doi.org/10.3389%2Ffonc.2021.759007
13. Zhao, Y., et al.: Predicting lymph node metastasis using histopathological images based on multiple instance learning with deep graph convolution. In: 2020 IEEE/CVF Conference on Computer Vision and Pattern Recognition (CVPR), pp. 4836–4845 (2020). https://doi.org/10.1109/CVPR42600.2020.00489
14. Zheng, X., et al.: Deep learning radiomics can predict axillary lymph node status in early-stage breast cancer. Nat. Commun. **11** (2020). https://doi.org/10.1038/s41467-020-15027-z
15. Zhou, L.Q., et al.: Lymph node metastasis prediction from primary breast cancer us images using deep learning. Radiology **294**(1), 19–28 (2020). https://doi.org/10.1148/radiol.2019190372, pMID: 31746687

A Client-Server Deep Federated Learning for Cross-Domain Surgical Image Segmentation

Ronast Subedi[5], Rebati Raman Gaire[5], Sharib Ali[4], Anh Nguyen[3], Danail Stoyanov[2], and Binod Bhattarai[1](✉)

[1] University of Aberdeen, Aberdeen, UK
binod.bhattarai@abdn.ac.uk
[2] University College London, London, UK
[3] University of Liverpool, Liverpool, UK
[4] University of Leeds, Leeds, UK
[5] NepAl Applied Mathematics and Informatics Institute (NAAMII), Lalitpur, Nepal

Abstract. This paper presents a solution to the cross-domain adaptation problem for 2D surgical image segmentation, explicitly considering the privacy protection of distributed datasets belonging to different centers. Deep learning architectures in medical image analysis necessitate extensive training data for better generalization. However, obtaining sufficient diagnostic and surgical data is still challenging, mainly due to the inherent cost of data curation and the need of experts for data annotation. Moreover, increased privacy and legal compliance concerns can make data sharing across clinical sites or regions difficult. Another ubiquitous challenge the medical datasets face is inevitable domain shifts among the collected data at the different centers. To this end, we propose a Client-server deep federated architecture for cross-domain adaptation. A server hosts a set of immutable parameters common to both the source and target domains. The clients consist of the respective domain-specific parameters and make requests to the server while learning their parameters and inferencing. We evaluate our framework in two benchmark datasets, demonstrating applicability in computer-assisted interventions for endoscopic polyp segmentation and diagnostic skin lesion detection and analysis. Our extensive quantitative and qualitative experiments demonstrate the superiority of the proposed method compared to competitive baseline and state-of-the-art methods. We will make the code available upon the paper's acceptance. Codes are available at: https://github.com/bhattarailab/federated-da.

Keywords: Domain Adaptation · Federated Learning · Decentralised Storage · Privacy

1 Introduction

The deployment of artificial intelligence (AI) technology in medical image analysis is rapidly growing, and training robust deep network architectures demands

B. Bhattarai et al. (Eds.): DEMI 2023, LNCS 14314, pp. 21–33, 2023.
https://doi.org/10.1007/978-3-031-44992-5_3

millions of annotated examples. Despite significant progress in establishing large-scale medical datasets, these are still limited in some clinical indications, especially in surgical data science and computer-assisted interventions [21]. Scaling training data needs multi-site collaboration and data sharing [1], which can be complex due to regulatory requirements (e.g. the EU General Data Protection Regulation [31], and China's cyber power [13]), privacy, and legal concerns. Additionally, even after training, practical AI model deployment in the clinic will require fine-tuning or optimization to local conditions and updates [11]. Therefore, architectures trained in federated and distributed ways to tackle cross-domain adaptation problems are critical. Yet, developing such architectures has challenges [25].

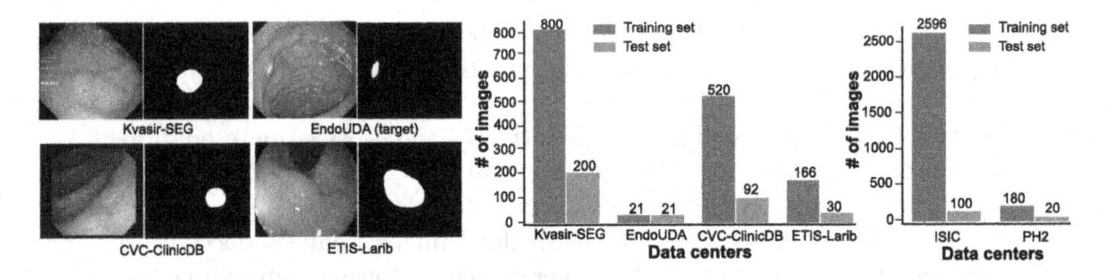

Fig. 1. Sample training examples collected from various centres for polyp segmentation (left); Sizes of training and test set at different centres for polyp segmentation (middle) and skin lesion segmentation (right).

Several works [7,8,29] have been proposed to tackle the problem of cross-domain adaptation in medical imaging. However, these methods require raw source and target domain data and cannot address the ever-increasing privacy concerns in sharing medical data. To circumvent the problem of privacy protection, there is a lot of research interest growing in Federated Learning (FL) in the medical domain [15,18,24,25,27,28,30]. Some methods even rely on synthetic data [10] to avoid sharing real data. For more details, we refer readers to a survey [23] on federated learning for smart health care. The common drawback of most existing methods [15,18,24,25] is that these methods are not designed for the domain shift problem. The most common topology in the FL workflow is averaging the local gradients (FedAvg) at the center and peer-to-peer gradient (FedP2P) sharing. These architectures are effective when data are independent and identically distributed (IID) in every client. In reality, domain shift is quite prevalent as data collected at different centers tend to be center specific. In Fig. 1, we can see the training examples for polyp segmentation collected at different centres. These examples show the discrepancy in lighting, camera pose and modalities in different centers. Some recent works, such as [9] by Guo et al. and [19] by Liu et al., address cross-domain problems in FL. However, [9] limits to a source-target pair at a time. Also, they employed adversarial loss to align the parameters, which is difficult to optimize. Similarly, FedDG [19] shares the information between the sources in the form of amplitudes of images. Their evaluation is limited to fundus images and MRI.

To tackle the problems of cross-domain adaptation and privacy protection in surgical image segmentation, we propose a simple yet effective Client-server FL architecture consisting of a server and multiple clients, as shown in Fig. 2. A server hosts a set of *immutable* task-specific parameters common to all the clients. Whereas every client requests the server to learn their domain-specific parameters locally and make the inference. In particular, every client learns an encoder's parameters to obtain an image's latent representation. These latent representations and ground truth masks are sent to the server. The decoder deployed on the server makes the predictions and computes the loss. The gradients are computed and updated only on the encoder to align the client's features with task-specific parameters hosted on the server. Aligning domain-specific parameters to common parameters helps diminish the gap between the source and target domains. We can draw an analogy between our framework and public-key cryptography. A client's network parameters are equivalent to a private key, and the decoder's parameters shared on the server are equivalent to the public key. Thus a client only with access to its private key can transfer its latent vector to the server containing the public key to obtain the semantic mask. Distributed storage of the parameters diminishes the risk of model parameter theft and adversarial attacks [20]. Moreover, each client communicates to the server only via a latent image representation, which prevents exposing the information of the raw data collected on the client side. It is possible to encrypt data transferred between the server and clients to secure communication. Finally, the server receives only fixed latent dimension representations, making it agnostic to the client's architecture. This enables clients to communicate with the server concurrently, improving efficiency. Likewise, none of the centres can modify the parameters deployed on the server; this would prevent the memorisation of client-specific information and parameter poisoning on the server [17].

To sum up, we propose a Client-server Federated Learning method for cross-domain surgical image segmentation. We applied our method to two multi-centre datasets for endoscopic polyp and skin lesion segmentation. We compared with multiple baselines, including recent works on cross-domain FL [9,19] and obtained a superior performance.

2 Method

Background: We consider a scenario where we have $C_1, C_2, \dots C_n$ represent n number of different institution's centres located at various geographical regions. Each centre collects its data in the form of tuple $(\mathbf{x}, \mathbf{y})$ where $\mathbf{x} \in \mathbb{R}^{w \times h \times c}, \mathbf{y} \in \mathbb{R}^{w \times h}$, where, w, h, c represent the width, height, and number of channels of an image. The annotated examples collected at the different centers are not IID due to variations in the illumination, the instruments used to acquire data, the ethnicity of the patients, the expertise of the clinician who collects the data, etc. We denote the total number of annotated pairs in each centre by N_n. In this paper, one of the major goals is to address the problem of domain adaptation, avoiding the need for the sharing of raw data to protect privacy.

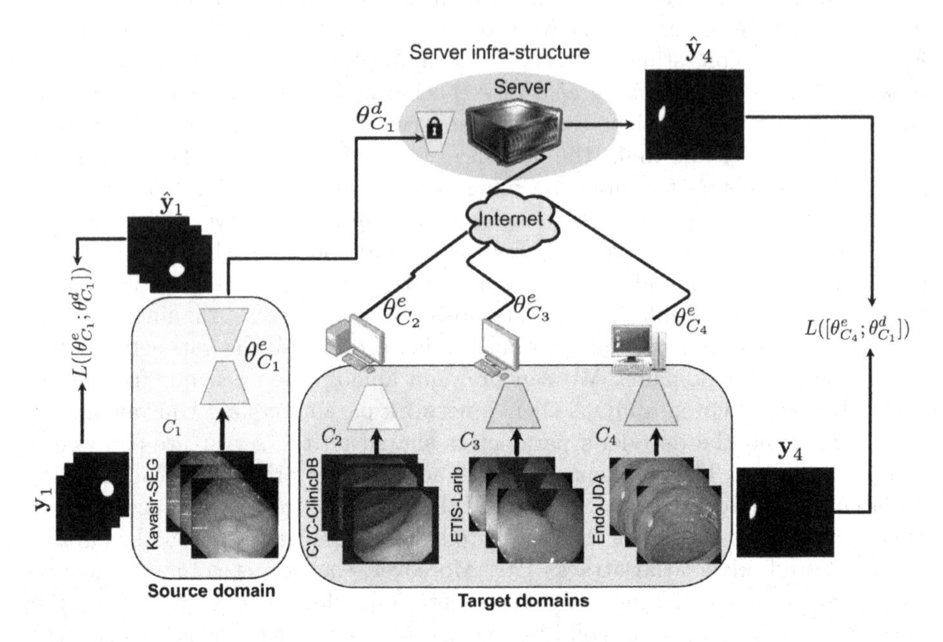

Fig. 2. The schematic diagram of the proposed framework. There are three major components: Source domain, Target domains, and Server infrastructure to share common parameters.

Learning Source Domain Parameters: First, we train a semantic segmentation network on the source data. In Fig. 2, the Source domain block shows the training of source domain/centre parameters. For us, data collected on Centre 1, C_1 is source data. We employ fully-convolutional encoder-decoder architecture. Such architectures are quite popular for semantic segmentation [2,26]. With the randomly initialised parameters, we minimise the objective of the Eq. 1. In Eq. 1, $\theta^e_{C_1}$ and $\theta^d_{C_1}$ represent the learnable parameters of the encoder and decoder, respectively.

$$L([\theta^e_{C_1};\theta^d_{C_1}]) = -\frac{1}{N_1}\sum_{i=1}^{N_1}\sum_{i=j}^{W}\sum_{k=1}^{H} \mathbf{y}_{ijk}\log\hat{\mathbf{y}}_{ijk} + (1-\mathbf{y}_{ijk})\log(1-\hat{\mathbf{y}}_{ijk}) \tag{1}$$

Setting-up Server Infra-structure: Figure 2's Server infra-structure block shows the setting up of the server infrastructure. Once we learn the parameters of the network from the source (C_1) data set, we upload the decoder's ($\theta^d_{C_1}$)

parameters on the server to share with every target client. The decoder module specializes to segment anatomies, given the encoder module's latent vector representation of the input image. As this segmentation task is common to all the centres, we propose to use a single decoder for all the centres. The previous works on cross-modal [4] and cross-feature [3] representations learning for cross-domain classification in computer vision employed the idea of sharing Convolutional Neural Networks' top layers parameters. However, none of these methods were employed in Federated Learning. The idea of sharing top layers parameters is in contrast to the conventional transfer learning [12] where the parameters are initialised with a model pre-trained on Imagenet and fine-tuned only the task-specific fully connected layers. We freeze the shared decoder parameters of the source. This arrangement brings advantages for privacy protection by preventing weight poisoning attacks [17]. Weight poisoning attacks alter the pre-trained model's parameters to create a backdoor. The parameters of the encoders can be shipped to the target client as per demand. Sometimes, the clients may demand these parameters to initialize their local networks when the training data is very small.

Federated Cross-domain Adaptation: Target centres other than the source centre deploy only the encoder network of their choice. In Fig. 2, the target domain block depicts the method. Every centre feeds its images to its encoder network during training, generating the respective latent representations. The latent representation and the ground truth ($\mathbf{y}_i$) mask from each target centre are pushed to the server where the pre-trained source decoder, θ^d_{C1}, is placed. The decoder feeds the latent representation, which predicts the output segmentation labels ($\hat{\mathbf{y}}_i$). We learn the parameters of the target encoders (θ^e_{Ci}) to minimize the objective given in Eq. 2. Since the decoder's parameters are frozen and shared with every client, only the target encoder's parameters are updated on the client side. This helps to align the latent representations to that of the source decoder's parameters and maximises the benefit from the task-specific discriminative representations learned from the large volume of source data.

$$\begin{aligned} L([\theta^e_{C_i};\theta^d_{C_1}]) = -\frac{1}{N_i}\sum_{i=2}^{N_i}\sum_{i=j}^{W}\sum_{k=1}^{H} \mathbf{y}_{ijk}\log\hat{\mathbf{y}}_{ijk} \\ +(1-\mathbf{y}_{ijk})\log(1-\hat{\mathbf{y}}_{ijk}) \\ \forall i \in 2,\ldots n \end{aligned} \quad (2)$$

The only thing that matters for target centres to communicate to the server is the fixed dimension of latent representations of an image. Thus, our architecture gives the flexibility of deploying the various sizes of networks on the client side based on available computing resources. And it is also entirely up to the target centres whether they want to initialize the parameters of the encoder using the parameters of the source domain. If the number of training examples is extremely few, then initialization using the pre-trained model's weight can prevent over-fitting.

Equations for Gradient Update: Equations 3 and 4 show mathematical formulation to update gradients of decoder and encoder modules in the proposed framework. Equation 3 updates the gradient of the decoder in the source center. Equation 4 represents the mechanism of updating gradients of each encoder module in every center. Here, α_i is the center-specific learning rate, where i represents the index of the centre. Please note the parameters of the decoder remain the same for every target centre. L denotes the loss function.

$$\theta^d_{C_1} = \theta^d_{C_1} - \alpha_1 \times \frac{\partial L([\theta^e_{C_1}; \theta^d_{C_1}])}{\partial \theta^d_{C_1}}; \tag{3}$$

$$\theta^e_{C_i} = \theta^e_{C_i} - \alpha_i \times \frac{\partial L([\theta^e_{C_i}; \theta^d_{C_1}])}{\partial \theta^e_{C_i}}; \forall i \in 1, \ldots n \tag{4}$$

3 Experiments

Data sets and Evaluation Protocol: We applied our method in two benchmark datasets: endoscopic polyp segmentation and skin lesion segmentation. The **polyp segmentation dataset** contains images collected at four different centres. Kvasir-SEG [14] data set is the source centre (C_1) in our experiment. It has 800 images in the train set and 200 in the test set. These high-resolution images acquired by electromagnetic imaging systems are made available to the public by Simula Lab in Norway. Similarly, the EndoUDA-target data set makes the first target domain (C_2) in our experiment, consisting of 21 images in both the training and testing sets [5]. Our experiment's second target domain centre (C_3) consists of images from the CVC-ClinicDB data set made available to the research community by a research team in Spain. There are 520 images in the train set and 92 in the test set. Finally, the ETIS-Larib data set released by a laboratory in France makes our third target domain data set (C_4). This data set consists of 166 in the train set and 30 images in the test set. These data sets were curated at different time frames in different geographical locations.

For **skin lesion segmentation**, we took data set collected at two different centres: ISIC (International Skin Imaging Collaboration) [6] and PH2 [22]. In ISIC, there are 2596 training examples and 102 test examples. The PH2 data set is curated through a joint research collaboration between the Universidade do Porto, Tecnico Lisboa, and the Dermatology Service of Hospital Pedro Hispano in Matosinhos, Portugal. In this data set, there are only 180 training examples and 20 testing examples. We consider ISIC and PH2 source and target domain, respectively. We report the mean Intersection over Union (mIoU) and dice scores for quantitative evaluations. Qualitative comparisons also validate our idea.

Baselines: We have compared the performance of our method with several competitive baselines, including both non-federated and federated frameworks. One of the naive baselines is to train a model for each target centre independently (*INDP*). The models of the with less training data overfit. Another configuration is creating a data pool by combining the training data (*COMB*) from all

Table 1. mIoU scores on Endoscopic Polyp Segmentation data sets (upper block) and Skin Lesion Segmentation data sets(lower block). The **AVERAGE** row is the weighted average across all the centers. The best score is highlighted in red and the second best is in blue.

Data	Centres	mIOU							
		INDP	COMB	[16]	[19]	FtDe	[9]	RandEn	FtEn
Endo.	Kvasir-SEG (C_1, source)	80.3	81.0	82.3	73.5	N/A	80.5	80.3	N/A
	EndoUDA (C_2)	52.0	57.5	53.1	29.7	59.9	61.7	50.6	62.0
	CVC-ClinicDB (C_3)	88.3	87.8	86.8	74.5	85.8	83.0	89.1	88.4
	ETIS-Larib (C_4)	62.1	66.9	61.4	70.8	65.1	71.7	64.3	69.9
	AVERAGE	**70.6**	**73.3**	**70.9**	**62.1**	**72.7**	**74.2**	**71.0**	**75.2**
Skin	ISIC (C_1, source)	81.3	75.7	84.9	N/A	N/A	N/A	81.3	N/A
	PH2 (C_2)	88.4	88.3	88.4	N/A	88.0	N/A	89.6	89.4

Table 2. Dice scores on Endoscopic Polyp Segmentation data sets (upper block) and Skin Lesion Segmentation data sets(lower block). The **AVERAGE** row is the weighted average across all the centers. The best score is highlighted in red and the second best in blue.

Data	Centres	Dice Score							
		INDP	COMB	[16]	[19]	FtDe	[9]	RandEn	FtEn
Endo.	Kvasir-SEG (C1, source)	89.1	89.5	**90.4**	84.7	N/A	89.2	89.1	N/A
	EndoUDA (C2)	64.8	73.0	0.694	45.8	74.9	76.3	67.2	76.5
	CVC-ClinicDB (C3)	93.8	93.5	92.9	85.4	92.4	90.7	94.2	93.8
	ETIS-Larib (C4)	76.6	80.2	76.1	82.9	78.9	83.5	78.3	82.3
	AVERAGE	**82.7**	**84.6**	**83.0**	**76.6**	**84.2**	**85.2**	**83.0**	**85.8**
Skin	ISIC (C1, source)]	89.7	86.2	91.8	N/A	N/A	N/A	89.7	N/A
	PH2 (C2)	93.8	93.7	93.8	N/A	93.6	N/A	94.5	94.4

the centres and training a single model. However, this method does not address any of the issues regarding privacy and compliance. Another viable option is to adapt a pre-trained model to a new domain by fine-tuning the parameters of the latter layers (*FtDe*). We also compared our method with competitive federated learning algorithms. FedAvg [16] averages the gradients computed in every center and shares the average gradients with the clients. This method ignores the non-IID nature of data from different centres. FedDG [19] is another Federated Learning method for domain adaption published at CVPR 2021. Finally, we also compared with another recent work by Guo et al. [9] for federated learning for multi-institutional data published at CVPR 2021. Our methods have two variants: initialising clients' parameters randomly(**RandEn**) and with the source's parameters (**FtEn**).

Implementation Details and Learning Behaviour: We implement our algorithms on PyTorch framework. All the images were resized to the dimension of 418×418. For optimization, we employ Adam optimizer with values of $\beta 1$ and $\beta 2$ set to 0.9 and 0.999 respectively. We initialize learning rate to 2e-4 and set the

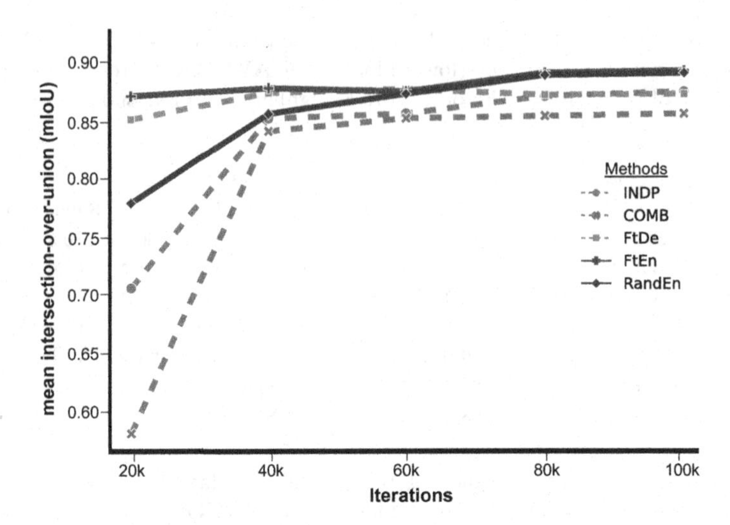

Fig. 3. Curves show the learning behaviour of different methods on PH2 data set.

decaying of the learning rate every 25k iterations. Taking UNet as a base architecture, we train the networks for 100k iterations and save the best-performing checkpoints on the validation set and report the performance on the test set. Since the task is to perform the semantic segmentation of the surgical images, we used Cross-Entropy loss to optimize the models. Figure 3 summarises the learning behaviour of the different methods for the first 100k iterations on PH2 data set, a target domain for skin lesion segmentation. The solid lines are our methods, and the dashed lines are the compared methods. The smooth curves demonstrate that our method, FtEn, has the highest starting mIoU. Hence, the method can be converged faster, and it is particularly beneficial in centers with limited computing resources for training.

Quantitative Evaluations: Table 1 shows the quantitative performance comparison. In the table, the last two grey-shaded columns show the performance of our methods. Our method outperforms *INDP* in every target centre. This signifies the importance of domain adaptation by our method. Compared to the other Federated Learning methods, our methods obtain the highest performance on 2/3 of target centres and are competitive on the third one for endoscopic polyp segmentation. On skin lesion segmentation, our method surpassed all the compared baselines and the recent competitive Federated Learning methods. Additionally, we also observe that our method FtEn, achieves the highest weighted average mIOU over all centres of polyp segmentation dataset. We also computed the dice score and obtained a similar performance, which is reported in Table 2.

Qualitative Evaluations: Figure 4 shows the qualitative performance comparisons between the baselines and the proposed methods on the target domains. Rows 2–4 (inclusive) are from endoscopy benchmarks, and the last row is from skin benchmarks. FedAvg fails to generalise well on target domains (see ETIS-

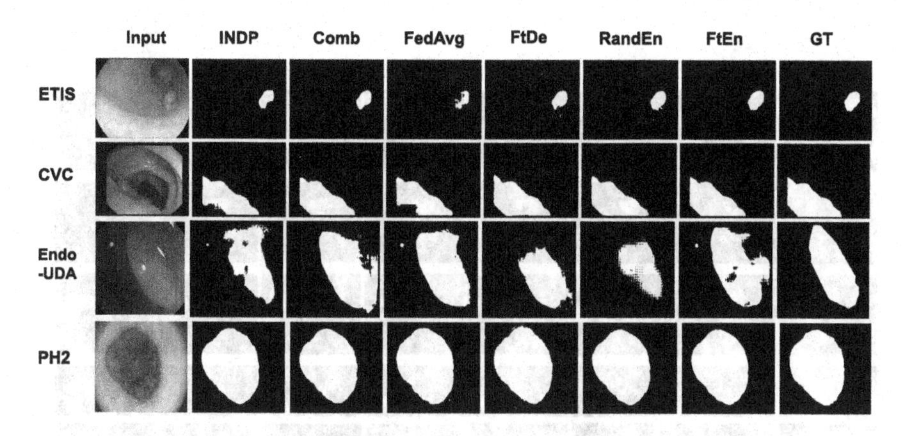

Fig. 4. Qualitative Comparisons

Larib and CVC-ClinicDB). Whereas our method is consistent in every target domain. These results further validate that our method is superior to the others. Figure 5 depicts additonal segmentation results on Endoscopic polyp data from the target domain centres. The first four rows are from the ETIS-Larib dataset, the next three are from the CVC-ClinicDB data set, and the last two are from EndoUDA. Columns **RandEn** and **FtEn** show the results of our method.

Computational Complexity: For INDP and FedAvg, the parameters of both the encoders and decoders grow in O(n) with the number of centres. Similarly, for FtDe and our method, the parameters of the encoder grow in O(n), while the parameters of the decoder are constant, i.e., O(1). Although the growth of the parameters for both the encoder and decoder for COMB is O(1), it does not address any privacy concerns. From these, our method is computationally less expensive and has high privacy protection.

Ablation Study on Varying Encoder Sizes: To further validate the robustness of our proposed framework, we trained the networks with different numbers of learnable parameters in the encoder module as shown in the Table 3. We vary the parameters by adding or removing the constituting layers in the encoding blocks of the network. We designate the conventional encoder of the UNet architecture as an encoder with medium size. The learnable parameters in the medium encoder are approximately 17 million. Table 4 depicts the architecture of a general encoder. It consists of three down-sampling layers, represented as *Down Block*. We vary the number of layers in the Down Blocks of the general encoder as per the availability of labeled data and computing resources in the particular centre.

The proposed federated distributed framework for domain adaptation provides flexibility to choose different network architectures to learn a common latent representation of images. These architectures can be designed based on the data size and available resources at a particular center. The performance evaluation results of various encoder sizes on polyp segmentation and skin lesion

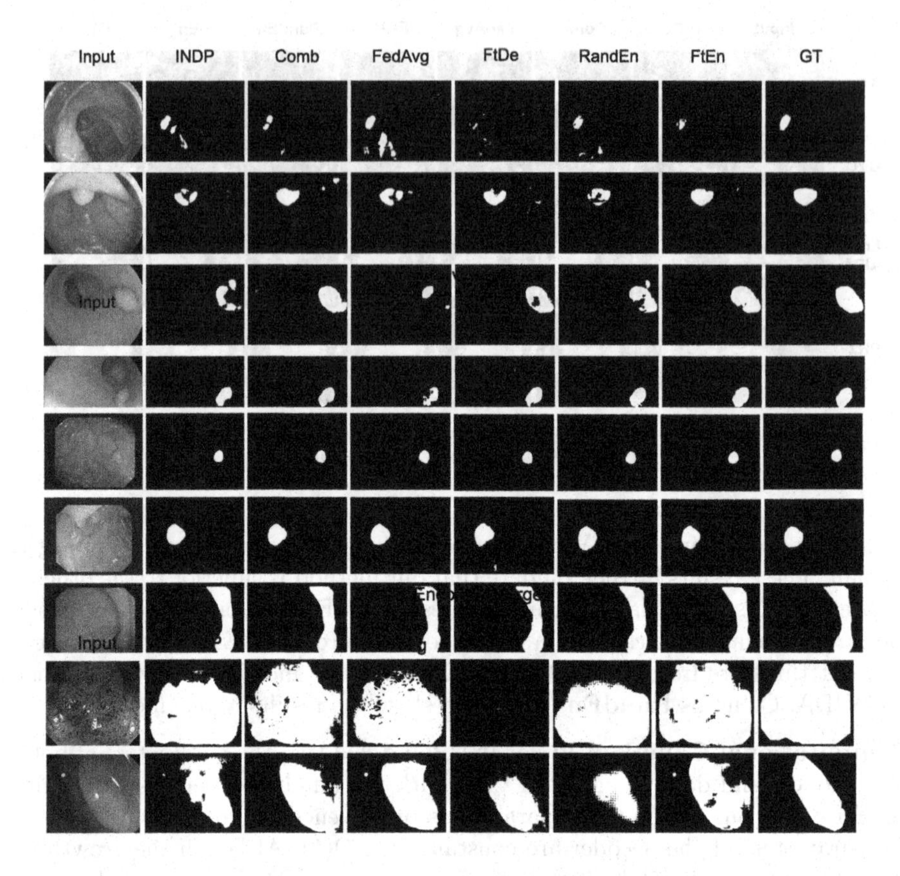

Fig. 5. Additional Qualitative Comparison oin Endoscopic Polyp Dataset

Table 3. Result on Endoscopic Polyp Segmentation Data sets (upper block) and Skin Lesion Segmentation (lower block) with different sizes of encoder networks.

Data	Centres	Size	Trainable Parameters	mIoU	
				INDP	RandEn(Ours)
Polyp	ETIS-Larib	Small	6,311,616	61.6	62.9
		Medium	17,080,896	62.1	64.3
		Large	27,850,176	62.5	64.7
Skin	PH2	Small	6,311,616	88.4	88.4
		Medium	17,080,896	88.5	89.6
		Large	27,850,176	89.4	89.8

segmentation are shown in Table 3. From the table, we can observe that segmentation performance for various sizes of the encoder is higher when trained using our framework than training independently for a specific center.

Table 4. Architecture of the general encoder.

Operations	Output Size
Input Image	$c \times h \times h$
Down Block	$c \times \frac{h}{2} \times \frac{h}{2}$
Down Block	$c \times \frac{h}{4} \times \frac{h}{4}$
Down Block	$c \times \frac{h}{8} \times \frac{h}{8}$

4 Conclusions

In this paper, we presented a client-server Federated Learning architecture for cross-domain surgical image segmentation. Our architecture addresses the cross-domain adaptation problem without sharing the raw images. Moreover, sharing only a part of the parameters from the source domain enhances privacy protection. Extensive experiments on two benchmarks from various data centres demonstrated improved cross-domain generalisation and privacy protection over the baselines and the competitive contemporary method.

Acknowledgements. This work is partly supported by the Wellcome/EPSRC Centre for Interventional and Surgical Sciences (WEISS) [203145Z/16/Z]; Engineering and Physical Sciences Research Council (EPSRC) [EP/P027938/1, EP/R004080/1, EP/P012841/1]; The Royal Academy of Engineering Chair in Emerging Technologies scheme; and the EndoMapper project by Horizon 2020 FET (GA 863146).

References

1. Ali, S., et al.: A multi-centre polyp detection and segmentation dataset for generalisability assessment. Sci. Data **10** (2023)
2. Badrinarayanan, V., Kendall, A., Cipolla, R.: Segnet: a deep convolutional encoder-decoder architecture for image segmentation. T-PAMI **39** (2017)
3. Bhattarai, B., Sharma, G., Jurie, F.: Deep fusion of visual signatures for client-server facial analysis. In: Proceedings of the Tenth Indian Conference on Computer Vision, Graphics and Image Processing, pp. 1–8 (2016)
4. Castrejon, L., Aytar, Y., Vondrick, C., Pirsiavash, H., Torralba, A.: Learning aligned cross-modal representations from weakly aligned data. In: Proceedings of the IEEE Conference on Computer Vision and Pattern Recognition, pp. 2940–2949 (2016)
5. Celik, N., Ali, S., Gupta, S., Braden, B., Rittscher, J.: EndoUDA: a modality independent segmentation approach for endoscopy imaging. In: de Bruijne, M., et al. (eds.) MICCAI 2021. LNCS, vol. 12903, pp. 303–312. Springer, Cham (2021). https://doi.org/10.1007/978-3-030-87199-4_29
6. Codella, N., et al.: Skin lesion analysis toward melanoma detection 2018: a challenge hosted by the international skin imaging collaboration (ISIC). arXiv preprint arXiv:1902.03368 (2019)

7. Ghafoorian, M., et al.: Transfer learning for domain adaptation in MRI: application in brain lesion segmentation. In: Descoteaux, M., Maier-Hein, L., Franz, A., Jannin, P., Collins, D.L., Duchesne, S. (eds.) MICCAI 2017. LNCS, vol. 10435, pp. 516–524. Springer, Cham (2017). https://doi.org/10.1007/978-3-319-66179-7_59
8. Gholami, A., et al.: A novel domain adaptation framework for medical image segmentation. In: Crimi, A., Bakas, S., Kuijf, H., Keyvan, F., Reyes, M., van Walsum, T. (eds.) BrainLes 2018. LNCS, vol. 11384, pp. 289–298. Springer, Cham (2019). https://doi.org/10.1007/978-3-030-11726-9_26
9. Guo, P., Wang, P., Zhou, J., Jiang, S., Patel, V.M.: Multi-institutional collaborations for improving deep learning-based magnetic resonance image reconstruction using federated learning. In: CVPR (2021)
10. Hu, S., Goetz, J., Malik, K., Zhan, H., Liu, Z., Liu, Y.: Fedsynth: gradient compression via synthetic data in federated learning. arXiv preprint arXiv:2204.01273 (2022)
11. Hu, Y., Jacob, J., Parker, G.J., Hawkes, D.J., Hurst, J.R., Stoyanov, D.: The challenges of deploying artificial intelligence models in a rapidly evolving pandemic. Nat. Mach. Intell. (2020)
12. Huh, M., Agrawal, P., Efros, A.A.: What makes imagenet good for transfer learning? arXiv preprint arXiv:1608.08614 (2016)
13. Inkster, N.: China's Cyber Power. Routledge, Abingdon (2018)
14. Jha, D., et al.: Kvasir-SEG: a segmented polyp dataset. In: Ro, Y.M., et al. (eds.) MMM 2020. LNCS, vol. 11962, pp. 451–462. Springer, Cham (2020). https://doi.org/10.1007/978-3-030-37734-2_37
15. Karargyris, A., et al.: Medperf: open benchmarking platform for medical artificial intelligence using federated evaluation. arXiv preprint arXiv:2110.01406 (2021)
16. Konečnỳ, J., McMahan, B., Ramage, D.: Federated optimization: distributed optimization beyond the datacenter. arXiv preprint arXiv:1511.03575 (2015)
17. Kurita, K., Michel, P., Neubig, G.: Weight poisoning attacks on pretrained models. In: Proceedings of the 58th Annual Meeting of the Association for Computational Linguistics, pp. 2793–2806 (2020)
18. Li, T., Sahu, A.K., Talwalkar, A., Smith, V.: Federated learning: challenges, methods, and future directions. IEEE Signal Process. Mag. **37** (2020)
19. Liu, Q., Chen, C., Qin, J., Dou, Q., Heng, P.A.: FedDG: federated domain generalization on medical image segmentation via episodic learning in continuous frequency space. In: Proceedings of the IEEE/CVF Conference on Computer Vision and Pattern Recognition, pp. 1013–1023 (2021)
20. Ma, X., et al.: Understanding adversarial attacks on deep learning based medical image analysis systems. Pattern Recognit. **110**, 107332 (2021)
21. Maier-Hein, L., et al.: Surgical data science-from concepts toward clinical translation. Med. Image Anal. (2022)
22. Mendonça, T., Ferreira, P.M., Marques, J.S., Marcal, A.R., Rozeira, J.: PH 2-A dermoscopic image database for research and benchmarking. In: 2013 35th Annual International Conference of the IEEE Engineering in Medicine and Biology Society (EMBC), pp. 5437–5440. IEEE (2013)
23. Nguyen, D.C., et al.: Federated learning for smart healthcare: a survey. ACM Comput. Surv. (CSUR) **55** (2022)
24. Parekh, V.S., et al.: Cross-domain federated learning in medical imaging. arXiv preprint arXiv:2112.10001 (2021)
25. Rieke, N., et al.: The future of digital health with federated learning. NPJ Digit. Med. (2020)

26. Ronneberger, O., Fischer, P., Brox, T.: U-Net: convolutional networks for biomedical image segmentation. In: Navab, N., Hornegger, J., Wells, W.M., Frangi, A.F. (eds.) MICCAI 2015. LNCS, vol. 9351, pp. 234–241. Springer, Cham (2015). https://doi.org/10.1007/978-3-319-24574-4_28
27. Sheller, M.J., et al.: Federated learning in medicine: facilitating multi-institutional collaborations without sharing patient data. Sci. Rep. **10**(1), 1–12 (2020)
28. Shen, Y., Zhou, Y., Yu, L.: CD2-pFed: cyclic distillation-guided channel decoupling for model personalization in federated learning. In: CVPR (2022)
29. Swati, Z.N.K., et al.: Brain tumor classification for MR images using transfer learning and fine-tuning. Comput. Med. Imaging Graph. (2019)
30. Tramel, E.W.: Siloed federated learning for multi-centric histopathology datasets. In: Domain Adaptation and Representation Transfer, and Distributed and Collaborative Learning: Second MICCAI Workshop, DART 2020, and First MICCAI Workshop, DCL 2020, Held in Conjunction with MICCAI 2020, Lima, Peru (2020)
31. Voigt, P., von dem Bussche, A.: The EU General Data Protection Regulation (GDPR). A Practical Guide, 1st edn. Springer, Cham (2017). https://doi.org/10.1007/978-3-319-57959-7

Pre-training with Simulated Ultrasound Images for Breast Mass Segmentation and Classification

Michal Byra[1,2](✉), Ziemowit Klimonda[1], and Jerzy Litniewski[1]

[1] Institute of Fundamental Technological Research, Polish Academy of Sciences, Warsaw, Poland
mbyra@ippt.pan.pl
[2] RIKEN Center for Brain Science, Wako, Japan

Abstract. We investigate the usefulness of formula-driven supervised learning (FDSL) for breast ultrasound (US) image analysis. Medical data are usually too scarce to develop a better performing deep learning model from scratch. Transfer learning with networks pre-trained on ImageNet is commonly applied to address this problem. FDSL techniques have been recently investigated as an alternative solution to ImageNet based approaches. In the FDSL setting, networks for transfer learning applications are developed using large amounts of synthetic images generated with mathematical formulas, possibly taking into account the characteristics of the target data. In this work, we use Field II to develop a large synthetic dataset of 100 000 US images presenting different contour objects, as shape features play an important role in breast mass characterization in US. Synthetic data are utilized to pre-train the ResNet50 classification model and various variants of the U-Net segmentation network. Next, the pre-trained models are fine-tuned on breast mass US images. Our results demonstrate that the proposed FDSL approach can provide good performance with respect to breast mass classification and segmentation.

Keywords: breast cancer · deep learning · synthetic data · ultrasound

1 Introduction

Convolutional neural networks (CNNs) have achieved *state-of-the-art* results in tasks like medical image classification and segmentation. However, medical data are often too scarce to develop better performing deep learning models from scratch. Transfer learning with networks pre-trained on large datasets, such as ImageNet, is commonly used to overcome the issue of limited data [10]. Pre-trained ImageNet networks are publicly accessible in the widely used deep learning frameworks and can be easily utilized for transfer learning purposes. ImageNet models have been developed using supervised learning with annotations prepared by humans. Additionally, self-supervised learning methods have been

B. Bhattarai et al. (Eds.): DEMI 2023, LNCS 14314, pp. 34–45, 2023.
https://doi.org/10.1007/978-3-031-44992-5_4

extensively researched to develop generic ImageNet networks without the human annotations. Formula-driven supervised learning (FDSL) has been recently investigated as an alternative solution to supervise and self-supervise learning. In FDSL, neural networks are trained on large amounts of synthetic images generated through mathematical formulas, thus removing the time-consuming data collection, curation and labeling process. Synthetic data can also be generated using formulas that take into account the characteristics of the target medical data. Kataoka et al. demonstrated that large-scale networks pre-trained on images featuring artificially generated fractals can be used to successfully process ImageNet data [18]. In subsequent work, Kataoka et al. also trained deep networks using radial contour images [17].

In this work, we present a FDSL method designed for ultrasound (US) image analysis. Acquiring large volumes of medical US data retrospectively from hospitals is challenging, particularly for rare diseases. Moreover, significant time is often required to properly curate and annotate US data. Handling patient data presents additional ethical challenges compared to ImageNet like datasets. We create a large synthetic US image dataset (SUD) featuring automatically generated variable contours obtained with an algorithm that uses Bezier curves and Perlin noise. Field II numerical software is utilized to simulate US images based on the generated contour objects and specified scatterer fields [15]. We use the synthetic dataset to pre-train classification and segmentation networks. Next, we compare the transfer learning capabilities of the models pre-trained on simulated US data and ImageNet with respect to breast mass classification and segmentation. Our results demonstrate the potential of using formula generated synthetic US data for network pre-training and transfer learning purposes in US.

2 Related Work

As far as we know, synthetic US data have not been so far utilized to train large vision models, such as the ResNet50 CNN. Synthetic US data have been used for machine learning purposes in several papers, mainly in the context of quantitative US. Chen et al. used Field II to simulate raw radio-frequency (RF) US data (US data before US image reconstruction) to pre-train a model for microbubble localization [9,15]. Byra et al. developed a Siamese CNN for temperature monitoring of tissues using RF US signals simulated in Field II [8]. Simson et al. generated synthetic RF (RF) US data from numerical breast phantoms using k-Wave software to train a deep CNN for sound speed estimation [25,26]. Similarly, Jush et al. used RF signals generated with K-wave to train a network for speed of sound estimation in breast [16]. Kim et al. utilized synthetic US data to train a deep learning model for attenuation coefficient estimation in liver [19]. Koike et al. used K-wave to generate RF US data and develop a network for aberration correction [21]. Moreover, synthetic US data have been also utilized for pre-training of US image reconstruction networks [14].

3 Methods

3.1 Simulated Ultrasound Data

Our approach to synthetic US data generation consists of two steps, see Fig. 1. First, we generate a binary mask indicating the position of the simulated contour object. Shape of the mask is related to one of 100 categories, corresponding to a specific set of hyper-parameters governing the stochastic mask generation procedure. Second, Field II software is used to create a b-mode US image based on the generated binary mask.

Mask Generation. We used mathematical formulas to generate objects of variable shapes. However, in comparison to the previous studies, utilizing fractals [18] or objects made by superimposing polygons [17], we generated compact contour objects. Our method was based on a two step procedure. First, we generated an n-sided polygon with vertices subjected to random perturbations, with coordinates given as follows:

$$\begin{aligned} x(i) &= (1 + A\epsilon_x(i))\cos(2\pi i/n), \\ y(i) &= (1 + A\epsilon_y(i))\sin(2\pi i/n), \end{aligned} \tag{1}$$

where A stands for the amplitude of 2D Perlin noise sequence $\epsilon(i) = (\epsilon_x(i), \epsilon_y(i))$ with $i \in [0, 1, ..., n]$. Next, we used the polygon to fit a smooth Bezier curve. In the second step, we utilized the generated curve to create a binary mask. 2D Bezier curve was appropriately scaled to match specific long-to-short axis ratio (LSAR). We selected this shape parameter because it corresponds to a general morphological feature, which is considered to be important for breast mass differentiation [11]. Next, we rotated the contour by a random angle sampled uniformly from $[-\pi/2, \pi/2]$ and rendered the contour as a compact binary mask to an 512×512 image, with position of the mask selected randomly. Mask was scaled to occupy a pre-specified area percentage of the entire image.

Contour Categories. Given the above mask generation procedure, we selected the following hyper-parameter set (n, A, LSAR) to generate shape categories for FDSL. The number of the vertices n was selected from $\{50, 150, 250, 350, 450\}$,

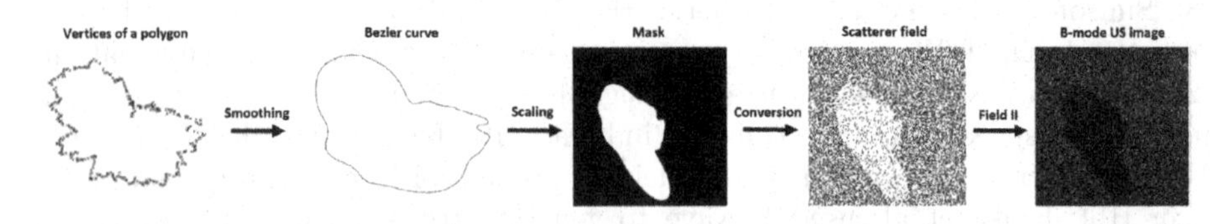

Fig. 1. Ultrasound image generation pipeline. First, a binary mask is generated based on a contour object obtained with a mathematical formula. Next, the binary mask in converted to a scatterer field and Field II simulation tool is used to generate a synthetic b-mode US image.

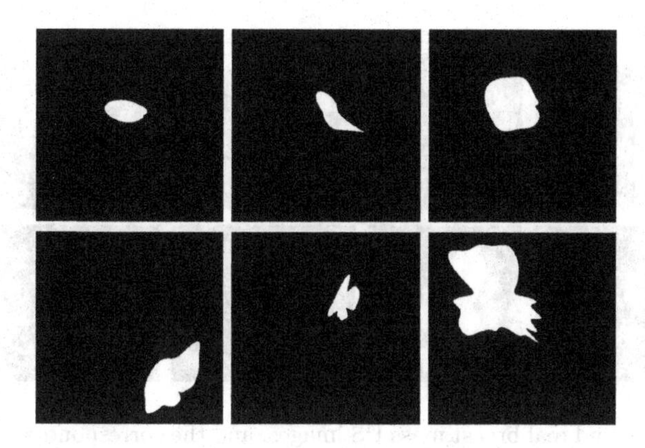

Fig. 2. Binary masks presenting contour objects of different variability.

Perlin noise amplitude A from {1, 5, 10, 20} and LSAR from {0.5, 3/4, 1, 4/3, 2}, respectively. The number of the vertices and noise amplitude levels were chosen based on our experiments and the study of Kataoka et al. [17]. The LSAR parameter was selected to span typical values of this parameter corresponding to benign and malignant breast masses. All combinations of these three parameters resulted in 100 categories, corresponding to objects with variable shapes. For each category, 1000 masks were generated. The masks occupied a percentage of the entire image that was sampled from {2%, 6.5%, 11%, 15.5%, 20%}. Figure 2 presents several images corresponding to contour objects of different variability.

Ultrasound Images. The Field II package was used to simulate b-mode US images based on the generated binary masks [15]. The medium was modeled as a cloud of point scatterers, with higher scatterer density corresponding to higher echogenicity in resulting b-mode image. Spatial distribution of the point scatterers was determined based on the binary mask. Area of the contour object was associated with a different scatter density than the background area. The background scatterer density was equal 267 mm^{-3}, which roughly corresponds to a single scatterer in a cube with edge length approximately equal to simulated pulse length. The scatterer density inside the object was equal to 0.2 or 5 of the background scatterer density. These values were experimentally selected to provide a satisfactory level of contrast in simulated b-mode US images. The scatterers locations were randomly drawn from the space extending from −15 mm to 15 mm in the x-axis (azimuth), −0.5 mm to 0.5 mm in y-axis (elevation), and from 10 mm to 40 mm in z-axis (depth). The medium with speed of sound equal to 1540 m m/s was assumed and the sampling frequency of 25 MHz was used. Linear probe with 128 elements, 0.298 mm pitch, 4 mm element height, 0.25 mm element width, 80% bandwidth and the transmit frequency 5 MHz was used in simulations. These parameters corresponded to a standard linear ultrasound probe. Synthetic US images were generated using fast plane wave imaging. In this setting, each US image was reconstructed based on a single transmit/receive

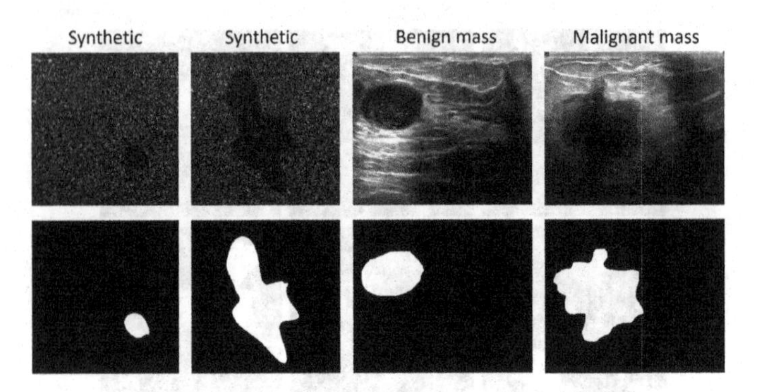

Fig. 3. Synthetic and real breast mass US images, and the corresponding binary masks.

event utilizing a plane wave emitted at angle equal to 0. We used this US imaging technique to simplify the calculations and generate the dataset in a fast manner. Still, about 30 d were required to generate the US images given our computing resources of several computers (e.g. AMD EPYC 7402P, Intel Xeon Processor E5-2680 V4 CPUs). Matlab and Parallel Computing Toolbox (MathWorks, MA, USA) were used for the simulations. Several simulated and regular breast mass US images are presented in Fig. 3. The generated dataset can be downloaded from the Zenodo repository (10.5281/zenodo.8196163).

3.2 Breast Mass Ultrasound Images

We used UDIAT and BUSI public datasets to assess the proposed FDSL approach [2,27]. Both datasets contain breast mass US images with binary labels and mass region annotations. For our experiments, we curated both datasets and removed duplicates and US images with scanner annotations overlaid over the breast mass region. This resulted in a total of 159 (52 malignant) US images from the UDIAT dataset and 400 (152 malignant) US images from the BUSI dataset. To evaluate the performance of the investigated deep learning methods, we applied stratified 5-fold cross-validation. The datasets were combined and divided into five training and test sets. Additionally, for each fold, we randomly selected 15% of cases from the training set to serve as a validation set. The proportion of UDIAT and BUSI images was the same for the training, validation, and test sets.

3.3 Classification

To demonstrate the usefulness of the proposed FDSL technique, we employed a standard approach to the classification of breast mass US images. Specifically, we used the ResNet50 CNN, which is the most widely used backbone model for transfer learning applications [13]. Residual networks have been utilized for breast mass classification in several papers [4,6,24].

We pre-trained the ResNet50 from scratch to predict 100 SUD categories. Given a synthetic US image, the network had to predict the category related to the underlying shape generation formula parameters. The architecture of the network was the same as in the original work, except for the last fully connected layer, which included 100 output units instead of 1000 as for ImageNet. Following the original paper and to enable comparisons with the ImageNet model, simulated US images were resized to 224×244 and scaled to $[-1,1]$ [13]. The network was trained for 250 epochs using the Adam optimizer with a learning rate set to 0.001 [20]. A cross-entropy loss function was used for the training, with a batch size set to 250. To augment the training set, we used the Albumentations package and applied the following operations with different probabilities: horizontal flipping, Gaussian filtering, median filtering, random brightness and contrast adjustments, and adaptive histogram equalization filtering [5]. In general, image texture filtering was applied with a probability of 75%. The experiments were performed using TensorFlow and NVIDIA 3090 RTX GPU [1].

Given pre-trained SUD model, two transfer learning techniques, commonly used in literature, were investigated to perform breast mass classification. First, we investigated if the SUD pre-trained model can serve as a good feature extractor [3]. In this case, the convolutional weights of the network were frozen and we replaced the last fully connected layer of the network with a single unit suitable for the binary classification task. Only the last linear layer, initialized with random weights, was trained to perform breast mass classification. For the second approach, we also replaced the last linear layer with a linear layer having a single output. However, in this case we did not freeze the weights and fine-tuned the entire network. Additionally, we investigated two standard approaches to the pre-processing of the input US images [3,6,12]. First, the network was trained with entire US images resized to 224×224. Second, we cropped the US images before the resizing based on manually segmented masks with a margin ranging from 10 to 30 pixels. This operation is commonly applied to exclude the noisy contributions from the surrounding tissues and improve the performance [3]. For the testing, we cropped the US images with a fixed margin of 20 pixels.

We evaluated the transfer learning performance with respect to the number of pre-training epochs on simulated US data. For each pre-trained model and each cross-validation fold, we used the training set of breast mass US images for fine-tuning. Adam optimizer with a learning rate of 0.001 was applied to fine-tune the model based on the binary cross-entropy loss function. During training, we monitored the loss function on the validation set and terminated training if no improvement was observed over 15 epochs. The same augmentation techniques were applied. For the training with the cropped US images, we performed augmentations before image cropping. For comparison, we investigated the ImageNet based ResNet50 as well as ResNet50 trained from scratch with the breast mass US images. We evaluated the performance of the models using accuracy and the area under the receiver operating characteristic curve (AUC). For the evaluations, we did not apply the filtering techniques to the breast mass US images.

Test scores were calculated for the models that achieved better performance on the validation set.

3.4 Segmentation with U-Nets

We also investigated if the simulated data can be used to pre-train breast mass segmentation networks. For this study, we examined three variants of the U-Net CNN [23]. We implemented the standard U-Net, Attention-U-Net and SK-U-Net [7,22]. Compared to the standard U-Net, Attention-U-Net utilizes an attention mechanism to process the feature maps propagated using the skip connections [22]. SK-U-Net utilizes selective kernel to adjust the receptive field of the network to better take into account that breast masses have variable sizes [7].

We pre-trained all three networks on simulated data for up to 100 epochs. The networks were pre-trained to predict the contour object mask based on synthetic US images. We used the Adam optimizer with a learning rate of 0.001 for the training. The masks and simulated US images were resized to 256×256. We used a loss function composed of the soft Dice score-based loss and binary cross-entropy loss, expressed with the following equation:

$$\mathcal{L}(S, A) = \mathcal{L}_{Dice}(S, A) + \alpha \mathcal{L}_{CE}(S, A), \quad (2)$$

where α stands for the cross-entropy loss weight set to 0.5, S and A are the reference mask and the network output, respectively. The same augmentation procedure was applied as in the case of the classification networks. Additionally, dropout regularization with a rate of 10% was applied to all convolutional layers of the networks. Following the pre-training, we fine-tuned each model with resized breast mass US images using the Adam optimizer and a learning rate of 0.0001. The same loss function was used for the fine-tuning. Moreover, for the comparison we also used the breast mass US images to train the models starting from random weights, following the same training procedure. Similarly to the classification models, at test time, the segmentation networks were evaluated with breast mass US images unprocessed with the image filtration techniques. To assess segmentation performance, we used the Dice score and detection rate metric defined as the percentage of cases with a Dice score above 0.5 [27]. Test scores were calculated for the models that achieved better performance according to the validation set.

4 Results

Figure 4 compares the weights of the first convolutional layer of the ResNet50 model pre-trained on the ImageNet and simulated US data. Training on the simulated data resulted in filters suited for the processing of grayscale images. Both networks presented filters sensitive to image texture variations. The network pre-trained for 100 epochs had filters with strictly localized spatial characteristics, which may be important for processing synthetic US images or anticipating early signs of benign overfitting.

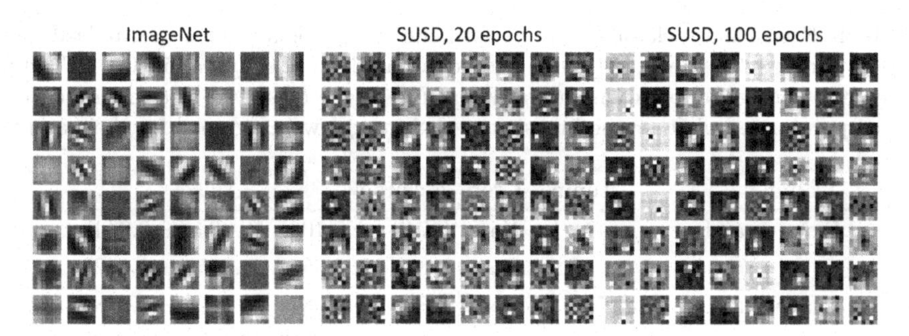

Fig. 4. Comparison between the weights of the first convolutional block of the ResNet50 pre-trained on ImageNet and simulated US data.

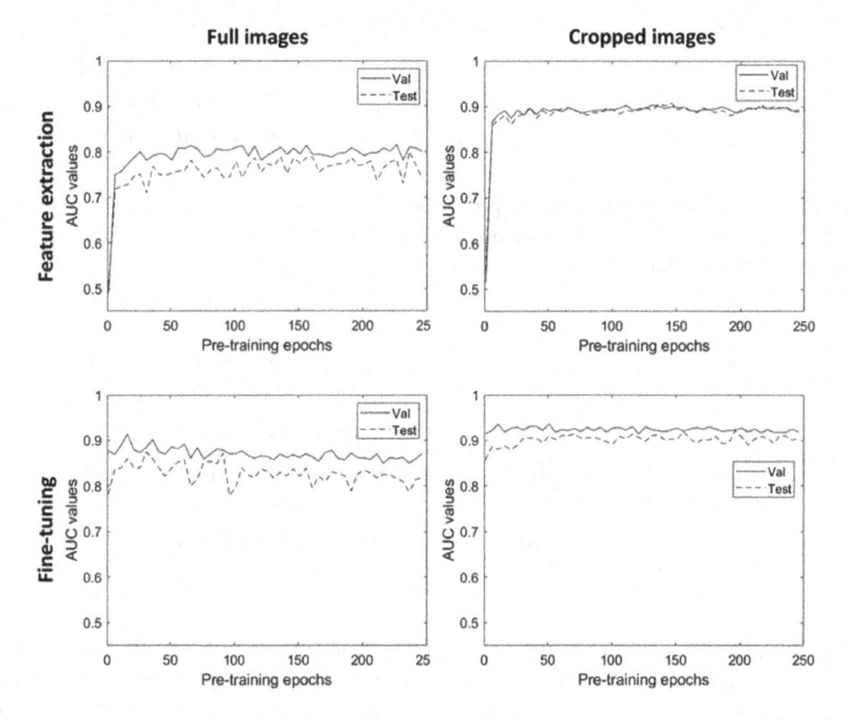

Fig. 5. Breast mass classification performance with respect to the number of pre-training epochs on synthetic data (starting from a single epoch).

Figure 5 presents breast mass classification performance with respect to the number of pre-training epochs on the synthetic data. We found that even a small number of the pre-training epochs was sufficient to achieve good classification performance. Table 1 presents the test set performance for each ResNet50 CNN selected based on the highest validation set AUC score. Training from scratch resulted in low AUC scores of 0.804 and 0.845 for full and cropped US images, respectively. All ImageNet models achieved good AUC scores of around 0.91. Models developed using cropped US images achieved comparable performance

Table 1. Breast mass classification performance test scores (mean, ± standard deviation) obtained for the investigated methods based on the ResNet50 CNN. Tuning indicates whether the backbone network was fine-tuned or served as a feature extractor, crop indicates whether the breast mass US images were cropped for the training.

Training	Tuning	Crop	Accuracy	AUC
From scratch	–	No	0.746 (± 0.109)	0.804 (± 0.044)
		Yes	0.801 (± 0.055)	0.845 (± 0.042)
ImageNet	No	No	0.831 (± 0.031)	0.909 (± 0.018)
		Yes	**0.884** (± 0.017)	**0.933** (± 0.023)
	Yes	No	0.854 (± 0.026)	0.927 (± 0.024)
		Yes	0.843 (± 0.031)	0.912 (± 0.017)
Simulated US data	No	No	0.752 (± 0.028)	0.783 (± 0.021)
		Yes	0.838 (± 0.044)	0.891 (± 0.021)
	Yes	No	0.813 (± 0.037)	0.859 (± 0.036)
		Yes	**0.852** (± 0.059)	**0.909** (± 0.032)

Table 2. Breast mass segmentation performance obtained for the investigated methods based on the U-Net CNN. Dice (mean, median, standard deviation) and DR stand for the Dice score and detection rate, respectively.

Network	Training	Dice	DR
U-Net	Scratch	0.608 (0.697 ± 0.304)	0.694
	Simulated US data	0.703 (0.849 ± 0.313)	0.796
SK-U-Net	Scratch	0.599 (0.701 ± 0.303)	0.699
	Simulated US data	**0.731** (0.866 ± 0.291)	**0.827**
Attention-U-Net	Scratch	0.529 (0.616 ± 0.331)	0.606
	Simulated US data	0.624 (0.794 ± 0.350)	0.706

to the ImageNet based networks, with AUC values of around 0.9. However, AUC values were much lower for SUD based models trained on full images. In this case, the model developed without fine-tuning achieved an AUC value of 0.783, similar to the ResNet50 trained from scratch. Fine-tuning of the model increased the AUC value to 0.859. The higher difference between SUD and ImageNet networks fine-tuned based on full images may be due to the lack of surrounding tissues in the simulated US images, which only presented contour objects.

Table 2 presents segmentation performance obtained for different variants of the U-Net CNN. Pre-training on the simulated data improved the performance of all investigated segmentation networks. For example, the mean Dice score obtained for the standard U-Net trained from scratch was equal to 0.608 (median 0.697) and increased to 0.709 (median 0.849) due to the SUD pre-training. The overall best performance was achieved by the SK-U-Net pre-trained on the simulated data, with a mean Dice score of 0.731 (median 0.866). Similarly, pre-training on the simulated data had a positive impact on the detection rates.

5 Discussion and Conclusion

Our results demonstrate the potential of the FDSL approach in US. We developed a dataset of 100 000 synthetic US images presenting various contour objects generated with a mathematical formula. Our findings confirm that models pre-trained with simulated data can be effectively fine-tuned for breast mass segmentation and classification, highlighting the potential of simulated US data for developing models suitable for clinical data. Notably, pre-training on the simulated US images improved performance compared to training from scratch, although the overall performance was comparable to that of ImageNet pre-trained networks. This may be due to the size and diversity of the ImageNet dataset, which includes over 1000k images across 1000 categories. The SUSD images presented single contour objects without any surrounding structures. However, our dataset of simulated images can be extended by incorporation of additional mathematical formulas and image generation methods to further improve the performance.

Our work has several limitations. First, generating large volumes of synthetic US data can be time and resource demanding. In our case, it took approximately 30 d to generate the synthetic US images using Field II. To speed up the simulations, we used fast plane wave imaging and limited the contents of the medium to contour objects of variable shape. To partially compensate for these simplifications, we applied image augmentation during the training of the networks. Similarly, Simson et al. used k-Wave package to generate pre-beamformed US data to train a network for sound speed estimation in breast [25]. Authors reported that it took approximately 43 d to simulate 5996 samples using plane wave imaging on a computer equipped with NVIDIA Quadro RTX 6000 GPU. Therefore, generating synthetic data for pre-training can be challenging. Second, our study focused only on breast mass segmentation and classification, and the usefulness of pre-training with synthetic US data for other US imaging tasks remains to be investigated. The performance improvement due to pre-training on synthetic data might depend on the similarity between the synthetic and clinical data distributions, which might differ between different imaging tasks. Thus, further research is needed to determine whether pre-training on synthetic US data is applicable to other US imaging tasks and how the amount and quality of synthetic data impact the performance of the pre-trained models. Third, our study used only one type of synthetic US data, and the impact of different types of synthetic data on pre-training needs to be explored. For example, it might be possible to generate synthetic US data that mimic different tissue types or imaging modalities.

In the future, we would like to utilize additional computing resources and generate a larger dataset of more detailed US images, including background objects corresponding to tissues of different physical properties. Simulations could be performed using k-Wave simulation tool, which takes into account the spatial distribution of sound speed and attenuation coefficients [26]. We consider our study as an important preliminary step towards generation of large-scale synthetic datasets for generic US image analysis. We believe that a large and diverse corpus of US datasets and pre-trained models will contribute to the development of robust and accurate machine learning methods for US image analysis.

Acknowledgments. The authors do not have any conflicts of interest. This work was supported by the National Science Center of Poland (2019/35/B/ST7/03792), program for Brain Mapping by Integrated Neurotechnologies for Disease Studies (Brain/MINDS) from the Japan Agency for Medical Research and Development AMED (JP15dm0207001) and the Japan Society for the Promotion of Science (JSPS, Fellowship PE21032).

References

1. Abadi, M., et al.: TensorFlow: large-scale machine learning on heterogeneous systems (2015). https://www.tensorflow.org/, software available from tensorflow.org
2. Al-Dhabyani, W., Gomaa, M., Khaled, H., Fahmy, A.: Dataset of breast ultrasound images. Data Brief **28**, 104863 (2020). https://doi.org/10.1016/j.dib.2019.104863
3. Antropova, N., Huynh, B.Q., Giger, M.L.: A deep feature fusion methodology for breast cancer diagnosis demonstrated on three imaging modality datasets. Med. Phys. **44**(10), 5162–5171 (2017)
4. Baccouche, A., Garcia-Zapirain, B., Elmaghraby, A.S.: An integrated framework for breast mass classification and diagnosis using stacked ensemble of residual neural networks. Sci. Rep. **12**(1), 1–17 (2022)
5. Buslaev, A., Iglovikov, V.I., Khvedchenya, E., Parinov, A., Druzhinin, M., Kalinin, A.A.: Albumentations: fast and flexible image augmentations. Information **11**(2), 125 (2020)
6. Byra, M.: Breast mass classification with transfer learning based on scaling of deep representations. Biomed. Signal Process. Control **69**, 102828 (2021)
7. Byra, M., et al.: Breast mass segmentation in ultrasound with selective kernel u-net convolutional neural network. Biomed. Signal Process. Control **61**, 102027 (2020)
8. Byra, M., Klimonda, Z., Kruglenko, E., Gambin, B.: Unsupervised deep learning based approach to temperature monitoring in focused ultrasound treatment. Ultrasonics **122**, 106689 (2022)
9. Chen, X., Lowerison, M.R., Dong, Z., Han, A., Song, P.: Deep learning-based microbubble localization for ultrasound localization microscopy. IEEE Trans. Ultrason. Ferroelectr. Freq. Control **69**(4), 1312–1325 (2022)
10. Deng, J., Dong, W., Socher, R., Li, L.J., Li, K., Fei-Fei, L.: Imagenet: a large-scale hierarchical image database. In: 2009 IEEE Conference on Computer Vision and Pattern Recognition, pp. 248–255. IEEE (2009)
11. Flores, W.G., de Albuquerque Pereira, W.C., Infantosi, A.F.C.: Improving classification performance of breast lesions on ultrasonography. Pattern Recogn. **48**(4), 1125–1136 (2015)
12. Han, S., et al.: A deep learning framework for supporting the classification of breast lesions in ultrasound images. Phys. Med. Biol. **62**(19), 7714 (2017)

13. He, K., Zhang, X., Ren, S., Sun, J.: Identity mappings in deep residual networks. In: Leibe, B., Matas, J., Sebe, N., Welling, M. (eds.) Computer Vision – ECCV 2016: 14th European Conference, Amsterdam, The Netherlands, October 11–14, 2016, Proceedings, Part IV, pp. 630–645. Springer, Cham (2016). https://doi.org/10.1007/978-3-319-46493-0_38
14. Hyun, D., et al.: Deep learning for ultrasound image formation: CUBDL evaluation framework and open datasets. IEEE Trans. Ultrason. Ferroelectr. Freq. Control **68**(12), 3466–3483 (2021)
15. Jensen, J.A., Svendsen, N.B.: Calculation of pressure fields from arbitrarily shaped, apodized, and excited ultrasound transducers. IEEE Trans. Ultrason. Ferroelectr. Freq. Control **39**(2), 262–267 (1992)
16. Jush, F.K., Biele, M., Dueppenbecker, P.M., Maier, A.: Deep learning for ultrasound speed-of-sound reconstruction: Impacts of training data diversity on stability and robustness. arXiv preprint arXiv:2202.01208 (2022)
17. Kataoka, H., et al.: Replacing labeled real-image datasets with auto-generated contours. In: Proceedings of the IEEE/CVF Conference on Computer Vision and Pattern Recognition, pp. 21232–21241 (2022)
18. Kataoka, H., et al.: Pre-training without natural images. In: Proceedings of the Asian Conference on Computer Vision (2020)
19. Kim, M.-G., Oh, S.H., Kim, Y., Kwon, H., Bae, H.-M.: Learning-based attenuation quantification in abdominal ultrasound. In: de Bruijne, M., Cattin, P.C., Cotin, S., Padoy, N., Speidel, S., Zheng, Y., Essert, C. (eds.) Medical Image Computing and Computer Assisted Intervention – MICCAI 2021: 24th International Conference, Strasbourg, France, September 27 – October 1, 2021, Proceedings, Part VII, pp. 14–23. Springer International Publishing, Cham (2021). https://doi.org/10.1007/978-3-030-87234-2_2
20. Kingma, D.P., Ba, J.: Adam: a method for stochastic optimization. arXiv preprint arXiv:1412.6980 (2014)
21. Koike, T., Tomii, N., Watanabe, Y., Azuma, T., Takagi, S.: Deep learning for hetero-homo conversion in channel-domain for phase aberration correction in ultrasound imaging. Ultrasonics **129**, 106890 (2023)
22. Oktay, O., et al.: Attention u-net: learning where to look for the pancreas. arXiv preprint arXiv:1804.03999 (2018)
23. Ronneberger, O., Fischer, P., Brox, T.: U-Net: convolutional networks for biomedical image segmentation. In: Navab, N., Hornegger, J., Wells, W.M., Frangi, A.F. (eds.) Medical Image Computing and Computer-Assisted Intervention – MICCAI 2015: 18th International Conference, Munich, Germany, October 5-9, 2015, Proceedings, Part III, pp. 234–241. Springer, Cham (2015). https://doi.org/10.1007/978-3-319-24574-4_28
24. Shen, Y., et al.: Artificial intelligence system reduces false-positive findings in the interpretation of breast ultrasound exams. Nat. Commun. **12**(1), 5645 (2021)
25. Simson, W.A., Paschali, M., Sideri-Lampretsa, V., Navab, N., Dahl, J.J.: Investigating pulse-echo sound speed estimation in breast ultrasound with deep learning. arXiv preprint arXiv:2302.03064 (2023)
26. Treeby, B.E., Cox, B.T.: k-wave: Matlab toolbox for the simulation and reconstruction of photoacoustic wave fields. J. Biomed. Opt. **15**(2), 021314–021314 (2010)
27. Yap, M.H., et al.: Automated breast ultrasound lesions detection using convolutional neural networks. IEEE J. Biomed. Health Inform. **22**(4), 1218–1226 (2018). https://doi.org/10.1109/JBHI.2017.2731873

Efficient Large Scale Medical Image Dataset Preparation for Machine Learning Applications

Stefan Denner[1,2(✉)], Jonas Scherer[1], Klaus Kades[1], Dimitrios Bounias[1,2], Philipp Schader[1], Lisa Kausch[1], Markus Bujotzek[1,2], Andreas Michael Bucher[5], Tobias Penzkofer[6], and Klaus Maier-Hein[1,3,4]

[1] Division of Medical Image Computing, German Cancer Research Center (DKFZ), Heidelberg, Germany

[2] Medical Faculty Heidelberg, University of Heidelberg, Heidelberg, Germany
`stefan.denner@dkfz-heidelberg.de`

[3] Pattern Analysis and Learning Group, Department of Radiation Oncology, Heidelberg University Hospital, Heidelberg, Germany

[4] National Center for Tumor Diseases (NCT), Heidelberg, Germany

[5] Institute for Diagnostic and Interventional Radiology, Goethe University Frankfurt, University Hospital, Frankfurt am Main, Germany

[6] Department of Radiology, Charité - Universitätsmedizin Berlin, Berlin, Germany

Abstract. In the rapidly evolving field of medical imaging, machine learning algorithms have become indispensable for enhancing diagnostic accuracy. However, the effectiveness of these algorithms is contingent upon the availability and organization of high-quality medical imaging datasets. Traditional Digital Imaging and Communications in Medicine (DICOM) data management systems are inadequate for handling the scale and complexity of data required to be facilitated in machine learning algorithms. This paper introduces an innovative data curation tool, developed as part of the Kaapana (https://github.com/kaapana/kaapana) open-source toolkit, aimed at streamlining the organization, management, and processing of large-scale medical imaging datasets. The tool is specifically tailored to meet the needs of radiologists and machine learning researchers. It incorporates advanced search, auto-annotation and efficient tagging functionalities for improved data curation. Additionally, the tool facilitates quality control and review, enabling researchers to validate image and segmentation quality in large datasets. It also plays a critical role in uncovering potential biases in datasets by aggregating and visualizing metadata, which is essential for developing robust machine learning models. Furthermore, Kaapana is integrated within the Radiological Cooperative Network (RACOON), a pioneering initiative aimed at creating a comprehensive national infrastructure for the aggregation, transmission, and consolidation of radiological data across all university clinics throughout Germany.

A supplementary video showcasing the tool's functionalities can be accessed at https://bit.ly/MICCAI-DEMI2023.

B. Bhattarai et al. (Eds.): DEMI 2023, LNCS 14314, pp. 46–55, 2023.
https://doi.org/10.1007/978-3-031-44992-5_5

Keywords: Medical Imaging · Data Curation · Machine Learning · Kaapana · Dataset Preperation · Quality Control · Bias Detection

1 Introduction

In recent years, the development and application of machine learning algorithms in medical imaging have emerged as an instrumental component in advancing healthcare and diagnostic accuracy [1]. This advancement, however, depends heavily on the availability and organization of high-quality medical imaging datasets [2,3]. The Digital Imaging and Communications in Medicine (DICOM) standard, commonly adopted for storing medical images, encapsulates both image data and vital metadata, including image modality, acquisition device manufacturer, and patient information like age and gender [4,5]. This metadata holds considerable value in the development of robust medical imaging machine learning algorithms [6]. Traditional DICOM data management systems, while effective for individual scans or patients, struggle to efficiently handle the scale and complexity of the data needed to be facilitated in machine learning algorithms [7]. The demand for superior data curation tools is crucial for advancing the field of medical imaging [8,9]. Despite recent progress in medical imaging data curation, existing solutions exhibit certain limitations. Some tools, while useful for data curation, are either proprietary, ill-equipped to handle large-scale medical datasets, or fail to fully exploit the benefits of DICOM headers [10].

Additionally, while there are automated approaches to enhance the data curation process, they are not conveniently integrated into a user-friendly tool [11–13].

In response to these challenges, we have developed an innovative data curation tool as part of the Kaapana open-source toolkit [14,15]. Kaapana is designed for advanced medical data analysis, especially in radiological and radiotherapeutic imaging, facilitating AI-driven workflows and federated learning approaches. By enabling on-site data processing and ensuring seamless integration with clinical IT infrastructures, it aims to address challenges in multi-center data acquisition and offers tools for standardized data processing workflows, distributed method development, and large-scale multi-center studies. Building up on Kaapana, our tool is designed to streamline the organization, management, and processing of large-scale medical imaging datasets, catering specifically to the needs of radiologists and machine learning researchers.

Our contribution is threefold: Our data management tool facilitates (1) efficient data curation by advanced search, auto-annotation and tagging, (2) quality control and review and (3) dataset bias detection by metadata visualization.

Kaapana is a constituent of the Radiological Cooperative Network (RACOON), an initiative to establish a nationwide infrastructure for collecting, transferring, and pooling radiological data across all German university clinics. Integrating our tool in Kaapana paves the way for its imminent deployment across all German university clinics, facilitating clinical validation.

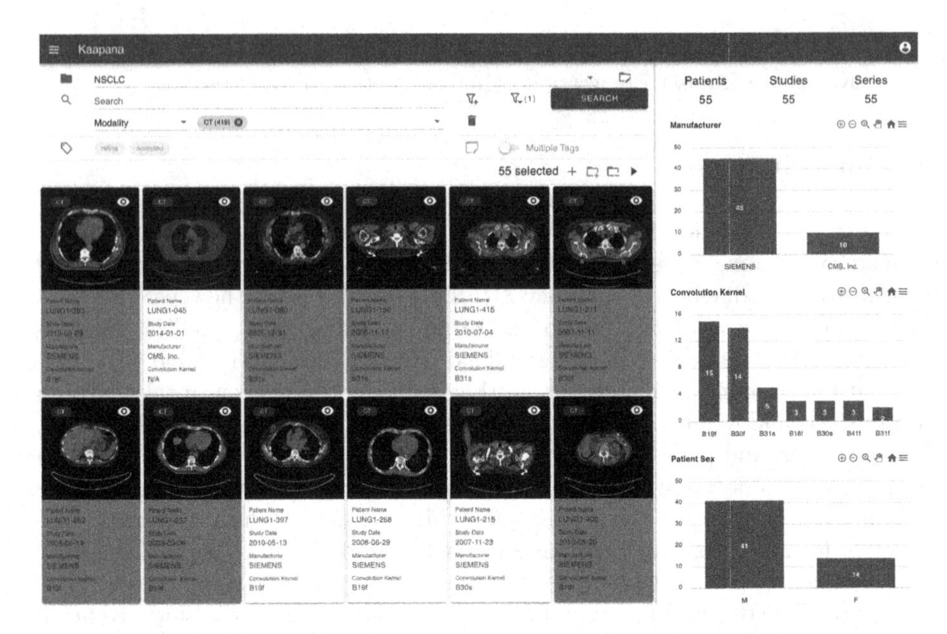

Fig. 1. Screenshot of the curation tool integrated into Kaapana. The gallery view displays series thumbnails accompanied by customizable metadata, providing a comprehensive visual overview. The sidebar showcases the metadata of the current selection, enabling swift detection of potential biases based on the DICOM metadata. This layout illustrates the tool's user-friendly interface and its capabilities in efficient data curation and bias detection.

2 Methodology

The essence of our methodology is to extend the capabilities of Kaapana, an open-source toolkit designed for medical data analysis and platform provisioning, by incorporating a comprehensive tool for managing, curating, and processing large-scale medical imaging datasets.

2.1 Technical Infrastructure

Our system benefits from Kaapana's robust technical infrastructure. Vue.js, a versatile JavaScript framework, powers the frontend, ensuring user-friendly, dynamic, and responsive web interfaces. FastAPI, a high-performance web framework, forms the backbone of the backend, enabling efficient communication with the frontend.

The persistence layer is three-fold, each serving a unique purpose. The dcm4chee Picture Archiving and Communication System (PACS) stores the original DICOM images, safeguarding their integrity and availability. For efficient management of large datasets, the DICOM Header is converted to JSON and stored in OpenSearch, a powerful open-source search engine known for its quick querying abilities. PostgreSQL, an open-source object-relational database

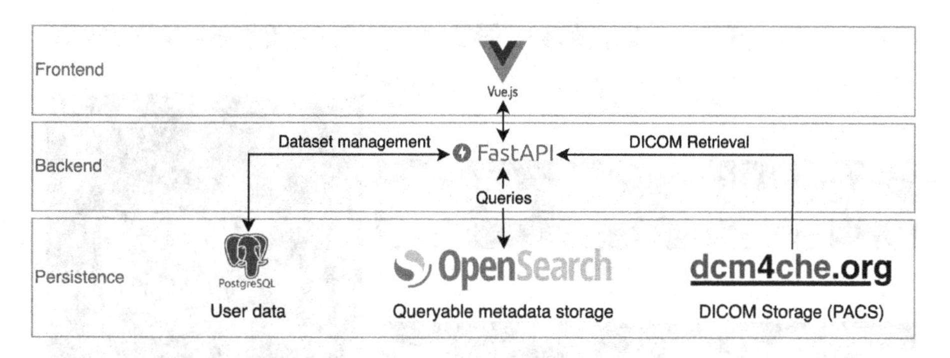

Fig. 2. Illustrating the technical infrastructure. Highlighting the frontend powered by Vue.js, the backend using FastAPI, and the three-fold data persistence layer consisting of dcm4chee, OpenSearch, and PostgreSQL. The arrows visualize the communication between the components.

system, forms the mapping layer, establishing connections between data and respective datasets, hence facilitating effective categorization and retrieval. The full utilized technical infrastructure is visualized in Fig. 2.

While our focus has primarily been on DICOM data, our solution also demonstrates flexibility in accommodating other formats. Kaapana is capable of transforming images in the Neuroimaging Informatics Technology Initiative (NIfTI) data format into DICOMs. These transformed images can then be curated. It's important to note, however, that metadata extraction is not possible from the NIfTI format; only the image data is preserved in the transformation. Nevertheless, this flexibility in data handling further extends the applicability of our tool in a variety of medical imaging contexts.

2.2 Graphical User Interface

The graphical user interface in seamlessly integrated into Kaapana's Vue.js frontend. Throughout the development process, which was conducted in close collaboration with radiologists, it was highlighted that varying use cases necessitate distinct user interface requirements. Consequently, the user interface has been designed to be highly adaptable, offering an array of customizable settings to cater to diverse needs. Overall, the user interface consists of a three-part layout visualized in Fig. 3.

Search. A sophisticated full-text search function, supporting wildcard search and free-text filtering, assists users in efficiently locating specific items based on image metadata. Additionally, it provides autocomplete functionality, streamlining the search process.

Gallery View. The gallery view provides a visual display of DICOM series, presenting them in a thumbnail format along with customizable metadata. The

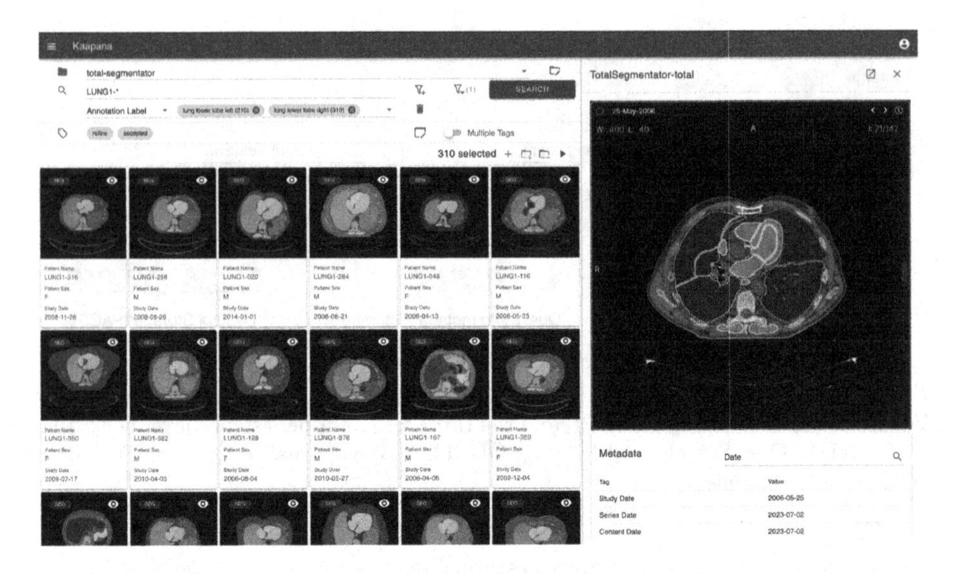

Fig. 3. Filtering for series containing lower lung lobes. The gallery view presents thumbnails with superimposed segmentations, while one selected series is opened in the sidebar for interactive 3D volume visualization with segmentations.

thumbnail creation is in compliance with the DICOM standard [4,5], which accommodates a broad spectrum of image modalities, including but not limited to, Structured Reports (SR), CT, or Magnetic Resonance Imaging (MRI). Given the current interest in segmentation algorithms within the medical imaging community [16], our tool automatically generates thumbnails for DICOM-SEGs or RTStructs that illustrates the segmentation superimposed on the original image. The gallery view also includes a multi-selection feature that facilitates bulk operations (see Fig. 1).

Sidebar. The sidebar serves a dual function - as a metadata dashboard and a detail view. The configurable metadata dashboard aggregates and displays comprehensive metadata distributions based on the current selection in the gallery view. These metadata distributions are interactive, allowing for selection and zooming for detailed examination. They can also be downloaded as charts or CSV files, providing flexibility in data analysis and sharing.

In the detail view mode, activated upon series selection, it showcases an interactive 3D visualization of the chosen DICOM series using the integrated (adjusted) OHIF Viewer [7] next to a searchable table with the series' metadata, including the DICOM Headers.

2.3 Machine Learning Integration

Kaapana is capable of executing state-of-the-art machine learning algorithms robustly. It is already equipped with a robust body part regression algorithm, allowing automatic assignment of which body part is covered in a given CT image [13,17]. Since one of Kaapana's major strengths is the easy extendability, we integrated TotalSegmentator [12] to even further extend the automatic data curation capabilities. TotalSegmentator is based on nnUNet [18], an automatically adapting semantic segmentation method, which allows segmenting 104 anatomical structures (27 organs, 59 bones, 10 muscles, 8 vessels) from CT images. This integration significantly enhances the automatic annotation capabilities. Furthermore, by this, users can filter for those body parts or anatomical structures and even further speed up their curation process.

2.4 Data Management and Workflow Execution

Our tool incorporates robust data management and workflow execution capabilities. Users can perform various actions on multiple selected series simultaneously, such as adding or removing series from a dataset and initiating workflows. An intuitive tagging system, with shortcut and autocomplete support, streamlines data annotation and categorization.

3 Results

Our data curation tool, integrated into Kaapana, provides a comprehensive and intuitive interface for managing, organizing, and processing extensive medical imaging datasets, thereby contributing significantly to efficient dataset curation for machine learning algorithms. Here, we highlight potential applications of our tool through a series of illustrative examples:

3.1 Dataset Management, Auto-Annotation and Tagging

Radiologists frequently handle vast collections of medical images, encompassing multiple patients, studies, and imaging modalities [9]. A common scenario involves a radiologist tasked with organizing thousands of CT and MRI scans acquired over several years for a large-scale study. Concurrently, in large-scale medical imaging studies curating and annotating an extensive collection of CT scans presents a formidable challenge. This requires a meticulous analysis of thousands of scans for visible disease symptoms, a process that is both labor-intensive and time-consuming.

Our tool offers a solution to these challenges with its gallery-style view, multi-select functionality, and advanced search features. Radiologists can swiftly sift through images, categorizing them into different datasets based on various attributes, such as patient demographics, study type, or imaging modality. The tool's advanced search functionality enables efficient image curation by allowing filters for DICOM metadata or algorithm outcomes, such as body part or anatomical structure.

These machine-assisted annotations provide an initial dataset that radiologists can validate and refine, significantly reducing the manual labor required and streamlining the annotation process. Furthermore, the gallery view, coupled with tagging functionality, enhances the organization of the curated and annotated dataset. This integrated approach to data organization, management, and annotation significantly alleviates the burden on radiologists and accelerates the preparation of data for machine learning applications.

3.2 Quality Control and Review

Our tool is particularly beneficial in scenarios where researchers need to validate the quality of images and segmentations in large medical imaging datasets, such as those obtained from multi-center studies. The tool's gallery and detail views can be effectively utilized to swiftly pinpoint images with poor quality or erroneous segmentations. An illustration of this capability is evident in the lower row of Fig. 4, where a multi-organ segmentation algorithm was applied to CT images but yielded subpar results. While 2D thumbnails may not always suffice for quality control of 3D segmentation algorithms, they can significantly expedite the quality control process in certain cases.

For instances where thumbnails fall short, the detail view allows researchers to navigate through the 3D volumes for a more comprehensive quality assurance. Moreover, as Magudia et al. [9] highlight, quality control for DICOM Headers is particularly crucial in multi-center studies due to data heterogeneity. Our tool caters to this need by displaying and allowing filtering of metadata.

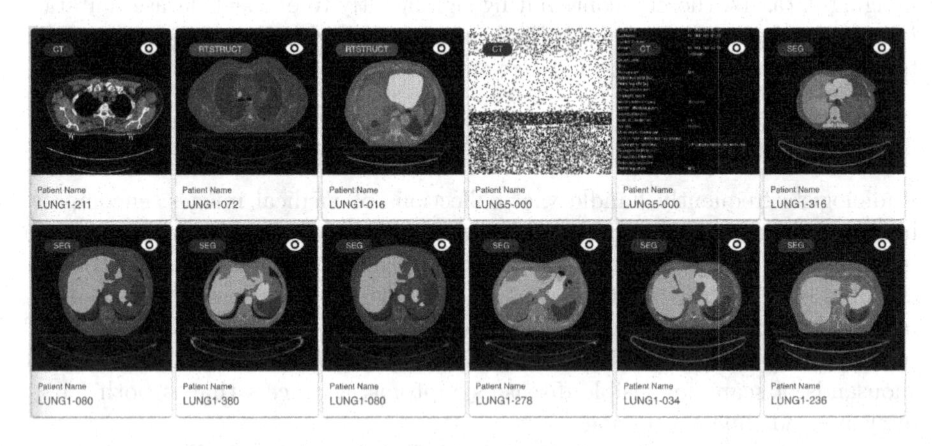

Fig. 4. Showcasing the gallery view's ability to handle various DICOMs and visually inspecting problematic series, such as the noisy series (top row, fourth from left) and the adjacent patient report. The radiologist can then exclude those problematic series. The lower row emphasizes the tool's capacity to quickly spot low-quality segmentations of a 3D CT image.

3.3 Uncovering Potential Bias in Datasets

Dataset biases, such as disparities in patient demographics or variations in scanner types and configurations, can profoundly influence the performance of machine learning models [6]. Such biases may lead to models that exhibit excellent performance during training and validation phases but falter in real-world applications due to an over-dependence on biased features. For example, a model predominantly trained on data from a specific scanner may struggle to generalize to images produced by other scanners [19]. Our tool can play a pivotal role in identifying these biases through its metadata dashboard. By aggregating and visualizing the metadata of selected items, researchers can discern patterns or inconsistencies that could signal potential biases. The visualization of the metadata distribution from a subset of the LIDC-IDRI Dataset's CT scans, as shown in Fig. 1, underscores the tool's ability to detect such biases [20]. A machine learning model trained on this dataset might inadvertently learn the skewed distribution of convolution kernels or scanners, which could result in failure on unseen data which does not represent the learned distribution.

By offering early detection of bias, the tool enables researchers to implement corrective strategies, such as data augmentation or bias mitigation techniques. This enhances the generalizability and resilience of the machine learning models developed, ensuring they perform optimally across varied scenarios.

4 Discussion and Conclusion

The development of an efficient data curation tool as part of the Kaapana open-source toolkit, as presented in this paper, addresses a critical need in the field of medical imaging. The availability and organization of high-quality medical imaging datasets are paramount for the successful application of machine learning algorithms in healthcare. The tool's integration with Kaapana provides a robust infrastructure for managing, curating, and processing large-scale medical imaging datasets.

One of the significant contributions of this tool is the streamlined annotation process. By employing advanced search functionality and auto-annotation capabilities through machine learning algorithms such as TotalSegmentator and Body Part Regression, the tool significantly reduces the manual labor required for image curation. Moreover, the tool's ability to support quality control and review mechanisms is vital for ensuring the reliability of datasets, especially in multi-center studies. The integration of a metadata dashboard is particularly noteworthy, as it enables the detection of potential biases in datasets. Furthermore, the open-source nature of the tool promotes collaboration and sharing among researchers, which is essential for advancing medical imaging research.

By leveraging Kaapana's federated learning capabilities, in future work curated datasets can be used in downstream federated learning use cases, enabling a collaborative approach to machine learning that respects data privacy and locality constraints. While the use cases demonstrate the utility of the tool, quantifying its enhancements remains a primary focus for future work.

Furthermore, integrating even more advanced algorithms for automatic image annotation could further improve the efficiency and accuracy of the tool. Another potentially promising advancement could be the integration of Electronic Health Record (EHR) data, which plays a crucial role in the process of creating datasets.

These future directions aim to ensure that the Kaapana data curation tool remains at the forefront of medical imaging research, catering to the evolving needs of radiologists and machine learning researchers.

Acknowledgments. Funded by "NUM 2.0" (FKZ: 01KX2121).

References

1. Anwar, S.M., Majid, M., Qayyum, A., Awais, M., Alnowami, M., Khan, M.K.: Medical image analysis using convolutional neural networks: a review. J. Med. Syst. **42**, 1–13 (2018)
2. Soffer, A., Ben-Cohen, A., Shimon, O., Amitai, M.M., Greenspan, H., Klang, E.: Convolutional neural networks for radiologic images: a radiologist's guide. Radiology **290**(3), 590–606 (2019)
3. DeGrave, A.J., Janizek, J.D., Lee, S.-I.: Ai for radiographic Covid-19 detection selects shortcuts over signal. Nat. Mach. Intell. **3**(7), 610–619 (2021)
4. Mildenberger, P., Eichelberg, M., Martin, E.: Introduction to the dicom standard. Eur. Radiol. **12**, 920–927 (2002)
5. Mustra, M., Delac, K., Grgic, M.: Overview of the dicom standard. In: 2008 50th International Symposium ELMAR, vol. 1, pp. 39–44. IEEE (2008)
6. Cruz, B.G.S., Bossa, M.N., Sölter, J., Husch, A.D.: Public covid-19 x-ray datasets and their impact on model bias-a systematic review of a significant problem. Med. Image Anal. **74**, 102225 (2021)
7. Ziegler, E., et al.: Open health imaging foundation viewer: an extensible open-source framework for building web-based imaging applications to support cancer research. JCO Clin. Cancer Inform. **4**, 336–345 (2020)
8. Willemink, M.J., et al.: Preparing medical imaging data for machine learning. Radiology **295**(1), 4–15 (2020)
9. Magudia, K., Bridge, C.P., Andriole, K.P., Rosenthal, M.H.: The trials and tribulations of assembling large medical imaging datasets for machine learning applications. J. Digital Imaging **34**, 1424–1429 (2021)
10. Diaz, O., et al.: Data preparation for artificial intelligence in medical imaging: a comprehensive guide to open-access platforms and tools. Physica Med. **83**, 25–37 (2021)
11. Nderitu, P., et al.: Automated image curation in diabetic retinopathy screening using deep learning. Sci. Rep. **12**(1), 11196 (2022)
12. Wasserthal, J., Meyer, M., Breit, H.-S., Cyriac, J., Yang, S., Segeroth, M.: Totalsegmentator: robust segmentation of 104 anatomical structures in ct images. arXiv preprint arXiv:2208.05868 (2022)
13. Tang, Y., et al.: Body part regression with self-supervision. IEEE Trans. Med. Imaging **40**(5), 1499–1507 (2021)
14. Scherer, J., et al.: Joint imaging platform for federated clinical data analytics. JCO Clin. Cancer Inform. **4**, 1027–1038 (2020)
15. Scherer, J., et al.: kaapana/kaapana: v0.2.0 (August 2022)

16. Maier-Hein, L., et al.: Why rankings of biomedical image analysis competitions should be interpreted with care. Nat. Commun. **9**(1), 5217 (2018)
17. Schuhegger, S.: Body part regression for ct images. arXiv preprint arXiv:2110.09148 (2021)
18. Isensee, F., Jaeger, P.F., Kohl, S.A.A., Petersen, J., Maier-Hein, K.H.: nnu-net: a self-configuring method for deep learning-based biomedical image segmentation. Nat. Methods **18**(2), 203–211 (2021)
19. Glocker, B., Robinson, R., Castro, D.C., Dou, Q., Konukoglu, E.: Machine learning with multi-site imaging data: an empirical study on the impact of scanner effects. arXiv preprint arXiv:1910.04597 (2019)
20. Armato, S.G., III., et al.: The lung image database consortium (lidc) and image database resource initiative (idri): a completed reference database of lung nodules on ct scans. Med. Phys. **38**(2), 915–931 (2011)

A Self-supervised Approach for Detecting the Edges of Haustral Folds in Colonoscopy Video

Wenyue Jin[1](✉), Rema Daher[2], Danail Stoyanov[2], and Francisco Vasconcelos[2]

[1] University College London, London, England
ucabwj0@ucl.ac.uk
[2] WEISS Centre, University College London, London, England

Abstract. Providing 3D navigation in colonoscopy can help decrease diagnostic miss rates in cancer screening by building a coverage map of the colon as the endoscope navigates the anatomy. However, this task is made challenging by the lack of discriminative localisation landmarks throughout the colon. While standard navigation techniques rely on sparse point landmarks or dense pixel registration, we propose edges as a more natural visual landmark to characterise the haustral folds of the colon anatomy. We propose a self-supervised methodology to train an edge detection method for colonoscopy imaging, demonstrating that it can effectively detect anatomy related edges while ignoring light reflection artifacts abundant in colonoscopy. We also propose a metric to evaluate the temporal consistency of estimated edges in the absence of real groundtruth. We demonstrate our results on video sequences from the public dataset HyperKvazir. Our code and pseudo-groundtruth edge labels are available at https://github.com/jwyhhh123/HaustralFold_Edge_Detector.

Keywords: Colonoscopy · Scene understanding · Edge detection · Landmark detection

1 Introduction

Reconstructing 3D gastrointestinal (GI) tract maps from endoscopy videos is a research challenge receiving increasing attention in recent years [4]. In the context of colon cancer screening, real-time 3D reconstruction would enable monitoring which surfaces have already been inspected [13,14], making it easier to ensure complete coverage and reduce the chance of missing polyps [18]. It would also enable complete reporting, associating polyps with precise colon map locations.

Simultaneous Localization and Mapping (vSLAM) is a popular algorithm framework that has been translated to colonoscopy 3D reconstruction [6,17].

Supplementary Information The online version contains supplementary material available at https://doi.org/10.1007/978-3-031-44992-5_6.

B. Bhattarai et al. (Eds.): DEMI 2023, LNCS 14314, pp. 56–66, 2023.
https://doi.org/10.1007/978-3-031-44992-5_6

However, we are still far from reliably reconstructing entire colons in real cases due to multiple imaging challenges. The majority of established methods builds 3D maps relying on the detection of point landmarks in the visualised scene across different frames. In colonoscopy, however, the detection and matching of point landmarks are extremely challenging due to scene textures being very similar, fast camera motions, abundant presence of light reflections, blur, and multiple types of occlusions.

While there is some research towards making point landmark detection more reliable in endoscopy [3], other alternatives involve bypassing the detection of points altogether. Some works perform registration of different frames by directly estimating depth [17] or relative motion [20] using end-to-end deep learning networks. The main challenge here is obtaining the necessary training data. Using virtual simulation has been suggested to train such algorithms [21], however, there is still a gap in generalising its results to real images.

A different alternative to bypass point landmark detection would be to focus on detecting scene edges instead. The colon anatomy has clearly visible and identifiable edges corresponding to its haustral folds (Fig. 1). While edge detection has seen significant progress in computer vision [19], there has been very little investigation on its application to endoscopy. Therefore, we introduce the following contributions:

- We introduce a method to detect haustral fold edges in colonoscopy based on the DexiNed architecture [19]. To the best of our knowledge, it's the first time this problem has been investigated.
- Given the inexistence of groundtruth for colonoscopy edge detection, we propose a combination of transfer learning and self-supervision to train our method.
- We propose an unsupervised evaluation process to measure the temporal consistency of edge predictions in continuous video frames.
- We will release both our code and pseudo-groundtruth edge masks for a subset of the public dataset HyperKvazir.

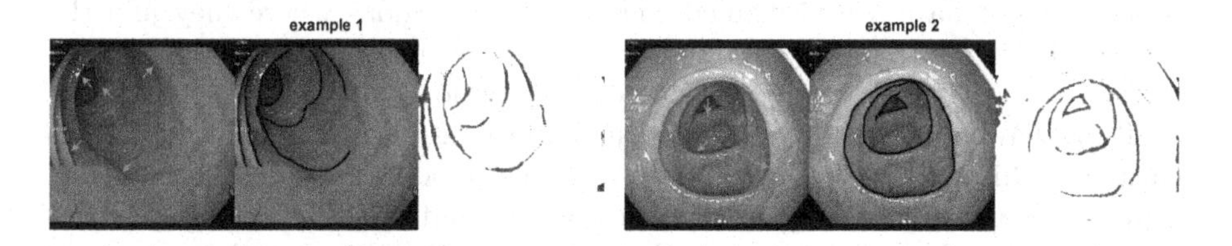

Fig. 1. We aim at detecting haustral folds (denoted by green arrows) in colonoscopy video. Formulating the problem as edge detection, these are circular contours on the colon wall (denoted by black lines). Sample predictions from our method are provided in black and white masks. (Color figure online)

2 Related Work

While most of the classic 3D navigation and reconstruction literature deals with rigid scenes containing unique and easy to recognise visual landmarks, in endoscopy there are two striking differences. The first one is the presence of deformable tissue. A few works have extended visual SLAM to explicitly model deformation of the 3D scene over time [12,22]. The second difference is that it is much more challenging to detect and track reliable landmarks on the GI tract due to simple tissue textures, frequent camera blur, light reflections and other dynamic occlusions. This paper will focus on this latter challenge which we now review in more detail.

Endoscopic scenes contain wet tissues illuminated by a close-range, moving light source. This produces abundant specular light reflections on tissue surfaces and makes it difficult to find landmarks with stable visual appearance. One approach to tackle this is to detect and inpaint specular reflections prior to landmark detection and matching [7]. A large amount of literature is dedicated to detection, filtering and inpainting of specular reflections in surgery [1,9,16].

The different visual appearance of endoscopic scenes presents a very specific domain shift in comparison to well established applications (e.g. outdoors/indoors human-made environments), and therefore machine learning approaches have been useful in bridging this gap. The SuperPoint [8] feature detector can be fine-tuned on endoscopy scenes in a self-supervised way [3], optimising its performance to this particular environment. There are a few other recent deep learning point feature detector alternatives that to the best of our knowledge have not been tried on endoscopy scenes [23,25,26].

Notably, there has been little investigation into the detection of features with other shapes than points. In the context of colonoscopy, this would be a promising direction since the colon is characterised by haustral folds, i. e. thin, ring-shaped structures on its surface (Fig. 1). A recent work has investigated the semantic segmentation of haustral folds [15]. However, we show that its results are still limited and inconsistent when applied to sequences of consecutive frames. We believe there is intrinsic ambiguity in labelling segmentations of these folds, as they do not have a well defined contour in the regions where they join the colon wall. Therefore, we propose to focus instead exclusively on the well defined portion of haustral fold contours using edge detection.

There have been recent advances in performing edge detection with deep learning architectures [19,24]. While pre-trained models are publicly available, these have been trained for general purpose vision, and we show in this paper that they are extremely sensitive to specular reflections. Furthermore, these methods have been trained in a fully supervised fashion, requiring either manually edge labels or proxy edges from semantic segmentation labels. While it would be a burdensome task to produce colonoscopy edge labels in sufficient numbers, we focus instead on self-supervised transfer learning.

3 Methodology

We aim at performing classification of each pixel in colonoscopic images as either edge or not-edge. Our target edges result from the colon shape (i.e. contours of the haustral folds) and not from its surface texture (i.e. vessels, shadows, reflections, etc).

Method Outline. As a baseline we start from the DexiNed model [19] which is a state-of-the-art edge detector trained on non-medical image data. This network is a sequence of 6 convolutional blocks, each of them performing pixel-dense edge detection at different image scales. Using skip connections and up-sampling, these 6 detections are fused into a final multi-scale edge detection result. The network is originally trained with a modified BDCN loss [10] in a fully supervised manner, using manually drawn edges as groundtruth labels.

A pre-trained model of DexiNed is publicly available, and we verify that it is able to detect haustral folds in colonoscopy videos. However, it also produces a significant amount of other false positive detections, mostly artifacts from illumination patterns. While DexiNed is pre-trained in a fully supervised manner, we aim at improving its results on endoscopy data without any additional groundtruth labels available.

Our first observation is that false positive detections can be removed by pre-processing the videos with a temporal specularity inpainting method [7]. In [7], a spatial-temporal transformer is used as a generator within a GAN structure to inpaint specular occlusions. While this produces very appealing results, unfortunately the pre-processing step restricts its usage to offline inference. This is because reliable inpainting results require processing a window of both past and future frames in a single inference step to take advantage of temporal cues. Furthermore such a pipeline would require running two different networks at inference time which is computationally sub-optimal.

To obtain a single model capable of online operation in an end-to-end fashion, we will leverage edges generated with offline pre-processing as pseudo-groundtruth labels to fine-tune DexiNed in a self-supervised manner.

Training Pipeline. Our training methodology is summarised in Fig. 2. We initialise the network with the weights from the original pre-trained DexiNed model, and then fine-tune it on endoscopy video. Our training procedure differs from [19] in the following aspects: (1) Instead of manually annotated groundtruth, we automatically generate pseudo-groundtruth labels with offline processing. (2) Instead of BDCN, we use a mean squared error (MSE) loss, as we empirically verified better results. (3) We train the network on batches of consecutive video frames rather than independent photos. (4) We also add a triplet loss term to improve temporal consistency in continuous video inference.

For a given set of training video clips $c = 1, ..., C$, we generate a set of pseudo-groundtruth label masks $G_{c,t}$ for all frames $X_{c,1}, ..., X_{c,T_c}$ in three steps. First, we pre-process all frames with the inpainting method from [7]. Secondly, we run

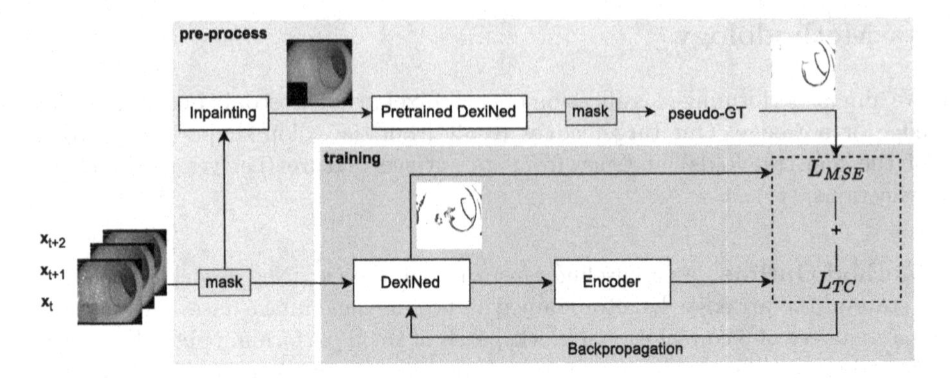

Fig. 2. Model fine-tuning. Pseudo-groundtruth labels are predicted by the pretrained DexiNed on the inpainted images in the pre-processing step. DexiNed is then further trained with a loss combining the pixel-wise loss L_{MSE} and consistency loss L_{TC}. The encoder uses a SegNet model to produce embedding vectors.

the pre-trained DexiNed model on the inpainted training data, generating clean edges of haustral folds. As a last step, we apply a mask to remove any edges resulting from field-of-view and interface overlays typically present in endoscopy images, resulting in the pseudo-groundtruth masks $G_{c,t}$.

Loss Function. We train our model with a loss combining two terms weighted by a parameter γ

$$L = \gamma L_{MSE} + (1-\gamma) L_{TC} \tag{1}$$

L_{MSE} is the mean squared error between edge predictions $E_{c,t}$ and respective pseudo-labels $G_{c,t}$

$$L_{MSE} = \frac{1}{P} \sum_{c=1}^{C} \sum_{t=1}^{T_c} \sum_{i=1}^{I} \sum_{j=1}^{J} (E_{c,t}(i,j) - G_{c,t}(i,j))^2 \tag{2}$$

where I, J are respectively the vertical and horizontal image resolution and P is the total number of pixels in the training data.

L_{TC} is a triplet loss that measures temporal consistency. We take edge predictions from 3 consecutive frames ($E_{c,t}$, $E_{c,t+1}$, $E_{c,t+2}$) and obtain their lower dimensional embedding vectors with an encoder $\psi()$. We use the encoder from SegNet [2], pre-trained on the Cars dataset[1] The triplet loss is then calculated:

$$L_{TC} = \sum_{c=1}^{C} \sum_{t=1}^{T_c-2} max(\|\psi(E_{c,t}) - \psi(E_{c,t+1})\|_2 - \|\psi(E_{c,t}) - \psi(E_{c,t+2})\|_2 + \beta, 0) \tag{3}$$

where β is a pre-defined margin parameter. In triplet loss terminology, $E_{c,t}$ represents the anchor, $E_{c,t+1}$ the positive sample, and $E_{c,t+2}$ the negative sample.

[1] The pretrained SegNet is available on https://github.com/foamliu/Autoencoder.

Evaluation. Edge predictions are typically evaluated by comparison against groundtruth labels, through Optimal Dataset Scale (ODS), Optimal Image Scale (OIS) and Average Precision (AP) [19,24]. In addition, our main motivation is to investigate edges as alternative features for navigation, and therefore we aim at temporally consistent estimations that can be further registered in video sequences for camera motion estimation. To this end, we also propose an unsupervised temporal consistency metric based on [27] that was originally introduced for semantic segmentation.

Our evaluation method is summarised in Fig. 3. We measure the temporal consistency TC_t^{t+1} of two independent edge predictions E_t, E_{t+1} by first warping E_t into E'_t using optical flow then measuring the overlap between E'_t and E_{t+1}.

We assume that edge predictions E'_t and E_{t+1} are binarised with a threshold T_1. For optical flow we use FlowNet 2.0 [11]. While [27] computes intersection over union (IoU) between E'_t and E_{t+1} we find this is not adequate for dealing with thin edges. Small spatial shifts in edge predictions result in drastic IoU decrease without it necessarily corresponding to a drastic decrease in edge consistency. Instead, we apply a distance transform to both $E'_{c,t}$ and $E_{c,t+1}$, generating grayscale fields with intensity values representing the distance to the closest edge. The distance fields are binarised with a threshold T_2, resulting in masks D'_t,D_{t+1} denoting all pixels with a distance smaller or equal to T_2 from edges in E'_t,E_{t+1} respectively. The temporal consistency TC_t^{t+1} is a class-weighted IoU between D'_t and D_{t+1}. We weight classes based on their frequency in the image, due to the extreme imbalance between edge and not-edge pixels. Finally, the metric is averaged on all pairs of consecutive frames in the test data.

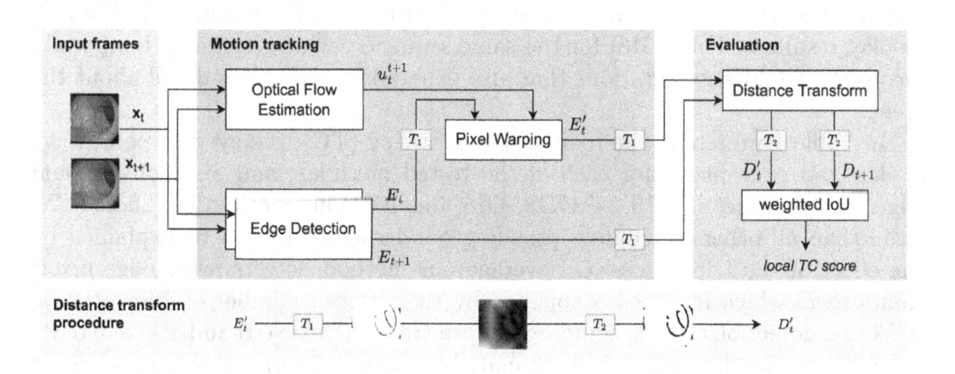

Fig. 3. The framework of consistency evaluation. The motion tracking block produces a pair of edge-maps E'_t and E_{t+1} aligned via optical flow (FlowNet 2.0). The overlap of aligned edge-maps is measured as the class-weighted IoU of binarised distance fields D'_t and D_{t+1}. These distance fields represent pixels within a distance T_2 to the edge predictions.

4 Experiments

Experimental Setup. We train and test our model on a subset of the Hyper-Kvasir dataset [5], defined as all 31 videos of lower GI with adequate bowel preparation (i.e. labelled as BBPS 2-3). We split the data into training, validation, and test with respectively 12, 8, and 11 videos. The images contain black margins and often an endoscope pose display on the lower left corner that produce irrelevant edge detections. We mask out these regions for all images. To compute temporal consistency metrics, we use the totality of the test video data. For comparison against groundtruth, we manually annotated a sparse sub-set of 78 randomly selected images from the test data.

Our method is implemented in Pytorch 1.12.1 with an Intel i7 CPU with 3 GHz and an Nvidia 3090 GPU. Video frames are cropped and resized to 256×256. DexiNed is trained with a RMSprop optimiser with $\alpha = 0.99$ and $\epsilon = 1 \times 10^{-8}$, using a constant learning rate $\eta = 1 \times 10^{-8}$. We use a triplet loss margin $\beta = 1$. A threshold $T_1 = 240$ is set to binarise edge-maps. We use $T_2 = 5$ for model evaluation. We used two-stage training where all models are trained with MSE loss for 5 epochs, followed by 5 epochs of our complete loss in Eq. 1.

Experimental Results Figure 4 displays qualitative results for our model, pseudo-groundtruth, and baselines for a sample sequence of 4 frames. Our model is able to significantly reduce the number of false positive detections caused by highlight reflections. This is a combined effect of the pseudo-groundtruth with temporal consistency (i.e. reflections are less consistent than haustral folds). We note that our method is able to capture the outer edge (see red box) which was not visible either in pre-trained DexiNed or pseudo-groundtruth. We also display results of Foldit [15] for the same sequence, which produces temporally inconsistent fold segmentations that also generally provide less detail about the scene.

In Table 1 we report the temporal consistency (TC), the average percentage of detected edge pixels for each of the tested methods, and also conventional edge accuracy metrics [19,24] ODS, OIS and AP. Our method has higher TC score than all others, including pseudo-groundtruth. This can be explained by the effect of the triplet loss. On average our method detects fewer edge pixels than others which in part is explained by the reduced number of false positive reflection detections (when compared to pre-trained DexiNed) and also due to its thinner edge predictions (when compared to pseudo-groundtruth). In terms of groundtruth evaluation, we observe a comparable performance to the pretrained model when evaluating edge detection metrics. The significantly higher scores obtained for the pseudo-groundtruth validate the reliability of our pseudo-labels. We also highlight that the FoldIt quantitative results should be interpreted with caution (we present them for the sake of completeness) as the detected regions are much larger than our proposed edges. However, its lower TC is consistent with the clearly visible temporal inconsistencies in Fig. 4. In Table 2, we show an ablation of the loss function weight γ. $\gamma = 1.0$ corresponds to using the MSE loss alone, which significantly reduces the TC score.

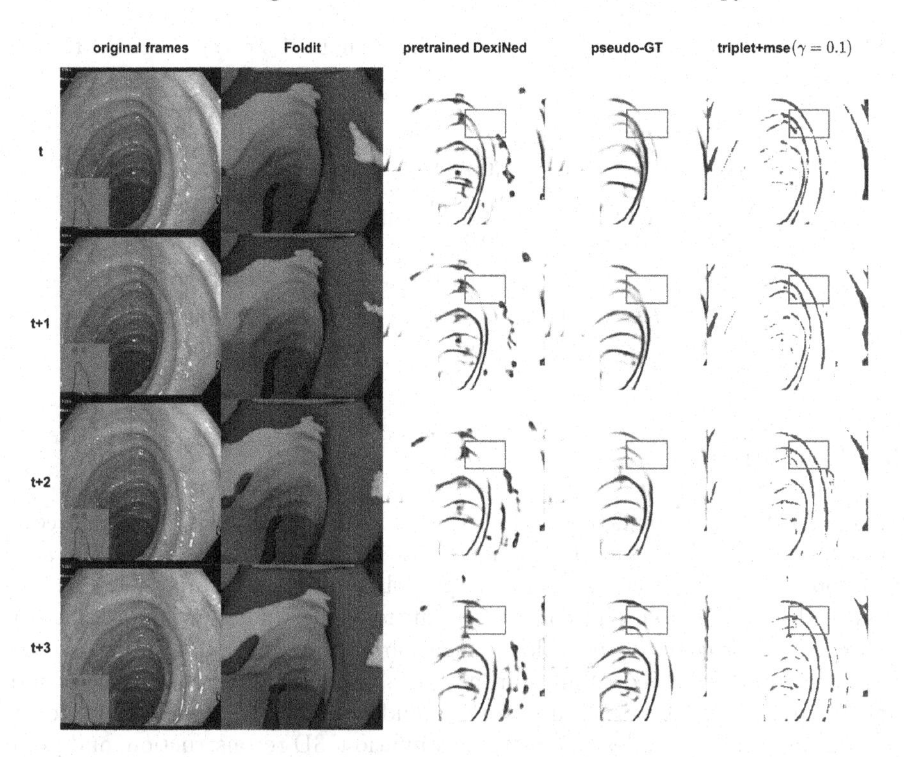

Fig. 4. Edge detection results on four consecutive frames. The predictions in the last column are made by our method. We highlight the red box where significant differences between methods and pseudo-groundtruth can be visualised. (Color figure online)

Table 1. Temporal Consistency (TC), edge pixel rate and results of edge detection metrics (ODS, OIS and AP). We note that Foldit is an image segmentation model (rather than edge detection), which explains the significant differences.

ID	Method	*TC* mean	*TC* std	edge pixel rate	ODS	OIS	AP
1	pretrained DexiNed	0.8840	0.0244	0.1564	0.6332	0.6613	0.5258
2	pseudo-GT	0.9028	0.0172	0.1300	**0.7271**	**0.7556**	**0.6704**
3	Ours (γ=.1)	**0.9348**	**0.0107**	0.0350	0.6491	0.6668	0.5145
4	Foldit	0.8708	0.0359	**0.4976**			

We must note that, as with any unsupervised metric, TC values cannot be analysed in a vacuum. In extreme, a method that never predicts any edge has the highest TC score but this is undesirable. Therefore we should also make sure edge pixel rates are not approaching zero. Our method has an edge pixel rate of 3.5% which is still deemed reasonable for the given data. We note that it is significantly lower than other methods due to detecting thinner edges.

Table 2. Ablation of loss weight γ. All values have similar TC except for $\gamma = 1$ (MSE).

ID	Method	*TC* mean	*TC* std	edge pixel rate
1	triplet+mse (γ=.1)	**0.9348**	**0.0107**	0.0350
2	triplet+mse (γ=.3)	0.9314	0.0111	0.0357
3	triplet+mse (γ=.5)	0.9278	0.0112	0.0351
4	triplet+mse (γ=.7)	0.9310	0.0110	0.0336
5	triplet+mse (γ=.9)	0.9268	0.0117	0.0409
6	triplet+mse (γ=1.0)	0.8445	0.0259	**0.4080**

5 Conclusions

We demonstrate that end-to-end detection of haustral fold edges in colonoscopy videos is feasible and can be made robust to the abundant reflection artifacts present in these scenes with a simple self-supervised training pipeline. We believe these are stable and consistent features across multiple views that can be exploited for colonoscopy video navigation and place recognition, but so far have been underexplored. While our method shows promising qualitative results and temporal consistency, future work should evaluate these features in downstream tasks such as endoscope motion estimation, 3D reconstruction, and place recognition.

Acknowledgments. This work was supported by the Wellcome/EPSRC Centre for Interventional and Surgical Sciences (WEISS) at UCL (203145Z/16/Z) and the H2020 FET EndoMapper project (GA863146). This work was partially carried out during the MSc in Robotics and Computation graduate degree at the Computer Science Department, UCL.

References

1. Ali, S., et al.: A deep learning framework for quality assessment and restoration in video endoscopy. Med. Image Anal. **68**, 101900 (2021)
2. Badrinarayanan, V., Kendall, A., Cipolla, R.: Segnet: a deep convolutional encoder-decoder architecture for image segmentation. CoRR abs/ arXiv: 1511.00561 (2015)
3. Barbed, O.L., Chadebecq, F., Morlana, J., Martínez-Montiel, J., Murillo, A.C.: Superpoint features in endoscopy. arXiv preprint arXiv:2203.04302 (2022)
4. Batlle, V.M., Montiel, J.M., Tardós, J.D.: Photometric single-view dense 3d reconstruction in endoscopy. In: 2022 IEEE/RSJ International Conference on Intelligent Robots and Systems (IROS), pp. 4904–4910. IEEE (2022)
5. Borgli, H., et al.: Hyperkvasir, a comprehensive multi-class image and video dataset for gastrointestinal endoscopy. Sci. Data **7**(1), 283 (2020). https://doi.org/10.1038/s41597-020-00622-y

6. Chen, R.J., Bobrow, T.L., Athey, T., Mahmood, F., Durr, N.J.: Slam endoscopy enhanced by adversarial depth prediction. arXiv preprint arXiv:1907.00283 (2019)
7. Daher, R., Vasconcelos, F., Stoyanov, D.: A temporal learning approach to inpainting endoscopic specularities and its effect on image correspondence (2022). https://doi.org/10.48550/ARXIV.2203.17013
8. DeTone, D., Malisiewicz, T., Rabinovich, A.: Superpoint: self-supervised interest point detection and description. In: Proceedings of the IEEE Conference on Computer Vision and Pattern Recognition Workshops, pp. 224–236 (2018)
9. Garcia-Vega, A., et al.: Multi-scale structural-aware exposure correction for endoscopic imaging. arXiv preprint arXiv:2210.15033 (2022)
10. He, J., Zhang, S., Yang, M., Shan, Y., Huang, T.: Bi-directional cascade network for perceptual edge detection. In: Proceedings of the IEEE/CVF Conference on Computer Vision and Pattern Recognition, pp. 3828–3837 (2019)
11. Ilg, E., Mayer, N., Saikia, T., Keuper, M., Dosovitskiy, A., Brox, T.: Flownet 2.0: evolution of optical flow estimation with deep networks. CoRR abs/ arXiv: 1612.01925 (2016)
12. Lamarca, J., Parashar, S., Bartoli, A., Montiel, J.: Defslam: tracking and mapping of deforming scenes from monocular sequences. IEEE Trans. Rob. **37**(1), 291–303 (2020)
13. Ma, R., et al.: Real-Time 3D reconstruction of colonoscopic surfaces for determining missing regions. In: Shen, D., et al. (eds.) MICCAI 2019. LNCS, vol. 11768, pp. 573–582. Springer, Cham (2019). https://doi.org/10.1007/978-3-030-32254-0_64
14. Ma, R., et al.: Rnnslam: reconstructing the 3d colon to visualize missing regions during a colonoscopy. Med. Image Anal. **72**, 102100 (2021)
15. Mathew, S., Nadeem, S., Kaufman, A.: FoldIt: haustral folds detection and segmentation in colonoscopy videos. In: de Bruijne, M., et al. (eds.) MICCAI 2021. LNCS, vol. 12903, pp. 221–230. Springer, Cham (2021). https://doi.org/10.1007/978-3-030-87199-4_21
16. Monkam, P., Wu, J., Lu, W., Shan, W., Chen, H., Zhai, Y.: Easyspec: automatic specular reflection detection and suppression from endoscopic images. IEEE Trans. Comput. Imaging **7**, 1031–1043 (2021)
17. Ozyoruk, K.B., et al.: Endoslam dataset and an unsupervised monocular visual odometry and depth estimation approach for endoscopic videos. Med. Image Anal. **71**, 102058 (2021)
18. Pickhardt, P.J., Taylor, A.J., Gopal, D.V.: Surface visualization at 3d endoluminal ct colonography: degree of coverage and implications for polyp detection. Gastroenterology **130**(6), 1582–1587 (2006)
19. Poma, X.S., Sappa, Á.D., Humanante, P., Akbarinia, A.: Dense extreme inception network for edge detection. CoRR abs/ arXiv: 2112.02250 (2021)
20. Rau, A., Bhattarai, B., Agapito, L., Stoyanov, D.: Bimodal camera pose prediction for endoscopy. arXiv preprint arXiv:2204.04968 (2022)
21. Rau, A., et al.: Implicit domain adaptation with conditional generative adversarial networks for depth prediction in endoscopy. Int. J. Comput. Assist. Radiol. Surg. **14**, 1167–1176 (2019)
22. Rodríguez, J.J.G., Montiel, J.M., Tardós, J.D.: Tracking monocular camera pose and deformation for slam inside the human body. In: 2022 IEEE/RSJ International Conference on Intelligent Robots and Systems (IROS), pp. 5278–5285. IEEE (2022)
23. Sarlin, P.E., DeTone, D., Malisiewicz, T., Rabinovich, A.: Superglue: learning feature matching with graph neural networks. In: Proceedings of the IEEE/CVF Conference on Computer Vision and Pattern Recognition, pp. 4938–4947 (2020)

24. Soria, X., Riba, E., Sappa, A.: Dense extreme inception network: towards a robust cnn model for edge detection. In: 2020 IEEE Winter Conference on Applications of Computer Vision (WACV), pp. 1912–1921. IEEE Computer Society, Los Alamitos, CA, USA (Mar 2020). https://doi.org/10.1109/WACV45572.2020.9093290, https://doi.ieeecomputersociety.org/10.1109/WACV45572.2020.9093290
25. Sun, J., Shen, Z., Wang, Y., Bao, H., Zhou, X.: Loftr: detector-free local feature matching with transformers. In: Proceedings of the IEEE/CVF Conference on Computer Vision and Pattern Recognition, pp. 8922–8931 (2021)
26. Tang, J., Ericson, L., Folkesson, J., Jensfelt, P.: Gcnv2: efficient correspondence prediction for real-time slam. IEEE Robotics Autom. Lett. **4**(4), 3505–3512 (2019)
27. Varghese, S., et al.: Unsupervised temporal consistency metric for video segmentation in highly-automated driving. In: 2020 IEEE/CVF Conference on Computer Vision and Pattern Recognition Workshops (CVPRW), pp. 1369–1378 (2020). https://doi.org/10.1109/CVPRW50498.2020.00176

Procedurally Generated Colonoscopy and Laparoscopy Data for Improved Model Training Performance

Thomas Dowrick(✉), Long Chen, João Ramalhinho, Juana González-Bueno Puyal, and Matthew J. Clarkson

Wellcome EPSRC Centre for Interventional and Surgical Sciences, UCL, London, England
t.dowrick@ucl.ac.uk

Abstract. The use of synthetic/simulated data can greatly improve model training performance, especially in areas such as image guided surgery, where real training data can be difficult to obtain, or of limited size. Procedural generation of data allows for large datasets to be rapidly generated and automatically labelled, while also randomising relevant parameters within the simulation to provide a wide variation in models and textures used in the scene.

A method for procedural generation of both textures and geometry for IGS data is presented, using Blender Shader Graphs and Geometry Nodes, with synthetic datasets used to pre-train models for polyp detection (YoloV7) and organ segmentation (UNet), with performance evaluated on open-source datasets.

Pre-training models with synthetic data significantly improves both model performance and generalisability (i.e. performance when evaluated on other datasets). Mean DICE score across all models for liver segmentation increased by 15% (p=0.02) after pre-training on synthetic data. For polyp detection, Precision increased by 11% (p=0.002), Recall by 9% (p=0.01), mAP@.5 by 10% (p=0.01) and mAP@[.5:95] by 8% (p-0.003).

All synthetic data, as well as examples of different Shader Graph/Geometry Node operations can be downloaded at https://doi.org/10.5522/04/23843904.

Keywords: Simulation · Image Guided Surgery · Data Engineering

1 Introduction

A majority of researchers in Image Guided Surgery (IGS) are involved with machine learning in some form (registration, segmentation, stereo reconstruction, classification etc.). However, the lack of application specific training data

B. Bhattarai et al. (Eds.): DEMI 2023, LNCS 14314, pp. 67–77, 2023.
https://doi.org/10.1007/978-3-031-44992-5_7

is a major blocker for development. In addition, the time-consuming process of manually labelling data is especially challenging for medical data, as labelling complex intraoperative scenes or radiological data typically requires the intervention of a trained clinician.

The wider computer vision community has benefited from large open-sourced datasets, including both real (e.g. ImageNet, KITTI) and simulated (e.g. Scene Flow, Virtual KITTI). While synthetic data can underperform in training when compared to real data, due to the so called 'domain gap', recent advances in domain specific simulation have produced models with equivalent, or in some cases superior performance to those trained only on real data [10,13,16]. Synthetic data has the additional advantage of being able to generate more accurate and complex labels, without the time/cost overheads of manual labelling.

IGS researchers have applied several methods from the machine learning community to generate synthetic data [4,6,9,11,14–16]. However, procedural generation, a method of creating data algorithmically from a pre-defined set of rules, is not an area that has had much investigation in IGS, despite showing promising results on conventional image recognition tasks [13]. As they allow control over the entire scene, procedural methods can also be integrated into active learning/active simulation pipelines, where the simulation parameters are updated on the fly in response to the network performance.

1.1 Contribution

In this work, methods for the generation of high quality, procedurally rendered data for IGS applications are described. This includes a fully procedural generation method, with no user inputs required, for generating colonoscopy data for polyp detection (which the authors believe to be a first), and a partially procedural method, where anatomically accurate models are used, with procedural textures, for liver segmentation during laparoscopic liver surgery.

Data was rendered using Blender (https://www.blender.org) (Fig. 1), taking ad-vantage of two main areas of functionality. The first is the use of Shader Graphs to generate realistic tissue textures. The second is Geometry Nodes, which allows for the entire geometry of the scene to be defined, and modified procedurally. On top of this, custom scripting allows for randomization of relevant parameters within the scene, allowing large, varied datasets to be rapidly generated.

The use of this data for model pre-training boosts performance, when evaluated on a number of publicly available datasets, in both laparoscopy and colonoscopy. Data used for training, along with original Blender files, to allow for data replication, is available for download (https://doi.org/10.5522/04/23843904).

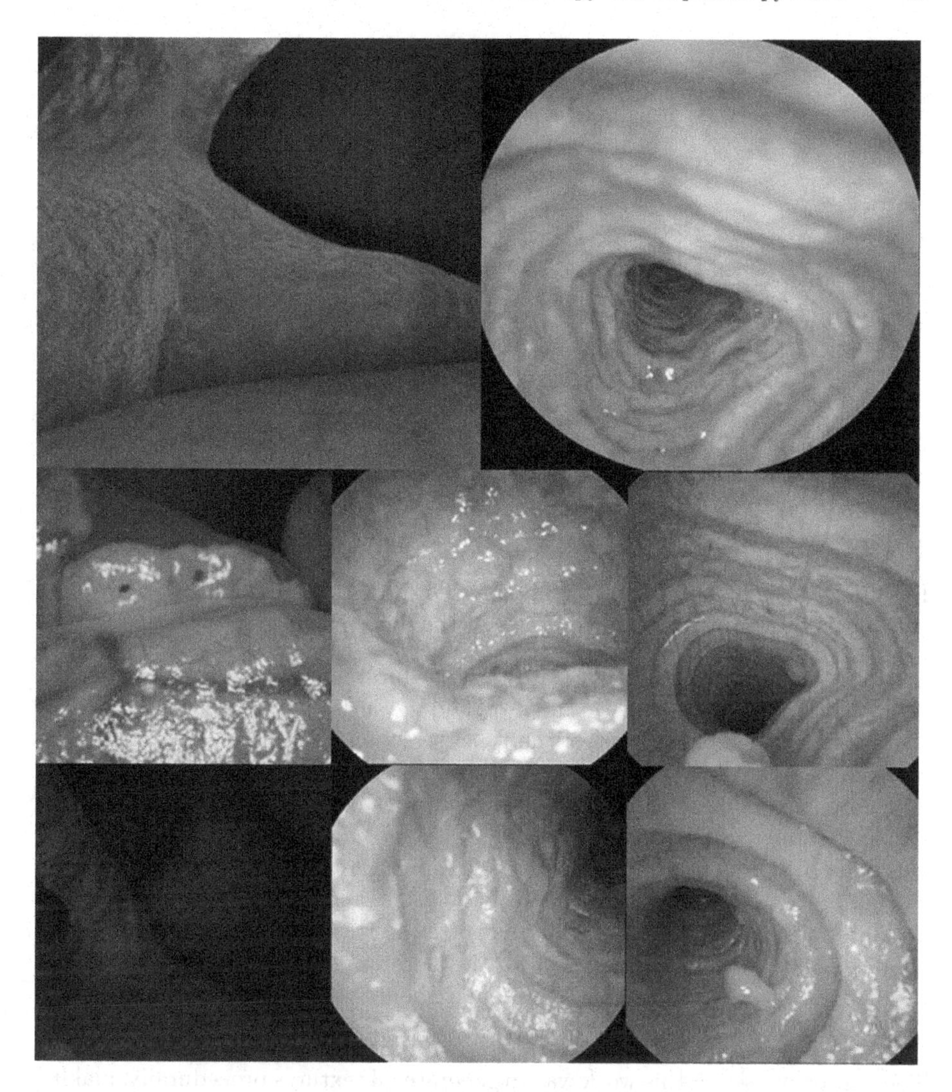

Fig. 1. Rendered synthetic data for Laparoscopy (left column) and Colonoscopy.

2 Methods

2.1 Shader Graphs for Texture Generation

The use of visual editors and node based approaches to producing shaders has increased in recent years, providing a layer of abstraction above shader code (HLSL, OSL etc.), allow the user to design shaders more intuitively, with instant feedback as parameters are changed. All major 3D graphics tools now include this functionality, including Unreal (Material Editor), Unity (Shader Graph),

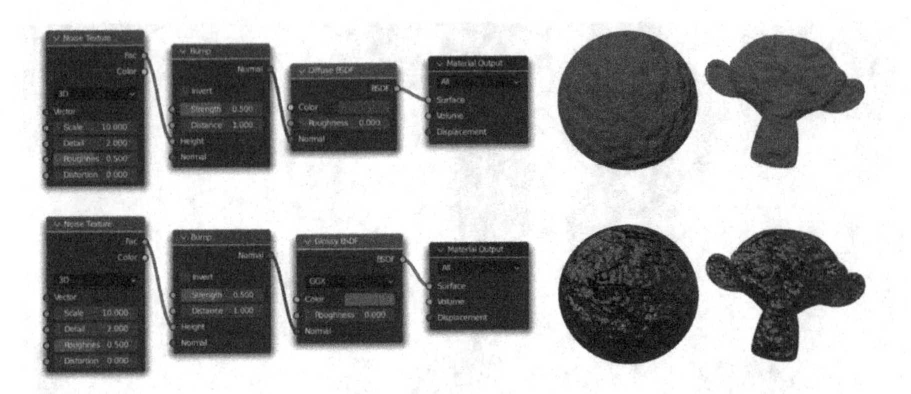

Fig. 2. Example of Blender Shader Graph for basic shading and normal maps. By combining different nodes, and adjusting their parameters, different textures/effects can be generated.

Houdini (Materials) and Blender. (Shader Graph) An example Shader Graph in Blender is shown in Fig. 2, making use of the following nodes:

- Noise Texture: this node generates a procedural noise pattern, often used for creating natural-looking textures or adding surface imperfections. It offers various parameters to control the type, scale, and intensity of the noise pattern.
- Bump: this node perturbs the surface normals of a material, simulating surface details without actually modifying the geometry.
- Diffuse BSDF: The Diffuse BSDF node represents a Lambertian diffuse material.
- Material Output: The Material Output node is the final node in the shader graph and serves as the endpoint for the material. It combines different shader outputs, such as Diffuse BSDF and links them to the surface of the 3D model for rendering.

The approach used in this work was to generate all textures procedurally, making use of more than 20 different Shader Graph nodes, allowing for fine grained control over all aspects of the texture's appearance, including albedo, bump mapping, displacement, subsur-face scattering, reflectance, glossiness etc.

Custom Shader Graphs were created for each organ, to match the properties of the real tissues as closely as possible. Within each graph, key parameters were identified which were to be randomly varied (Fig. 3) at simulation time, as well as the ranges over which to randomise. The appearance of each Shader Graph was manually tuned, and the appearance compared visually to sample images of each target tissue. All textures used for training in this work were generated using a Shader Graph, and these can be found in the accompanying dataset release (https://doi.org/10.5522/04/23843904).

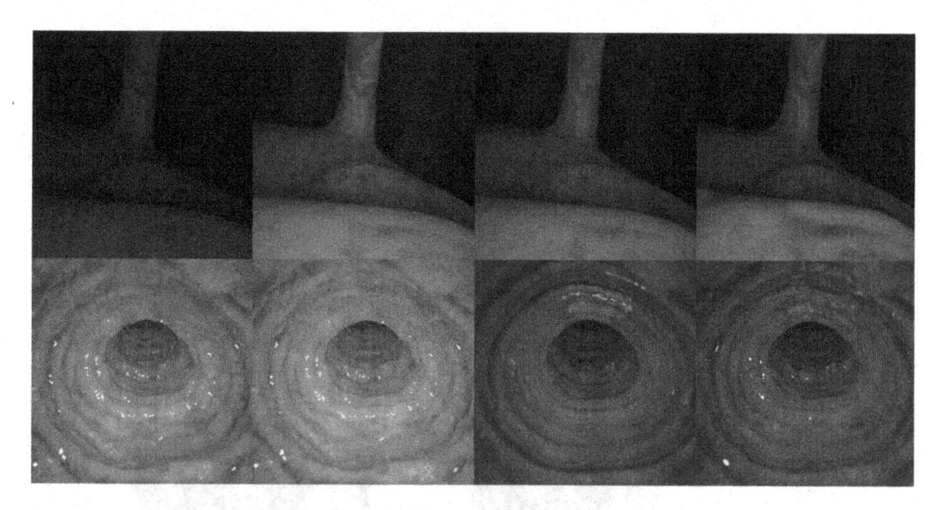

Fig. 3. Randomising texture and lighting parameters on laparoscopy (above) and colonos- copy (below) data.

2.2 Geometry Nodes for Model Generation

Procedural modeling is a powerful approach in computer graphics and 3D design that allows for the automatic generation of complex shapes, textures, and animations using algorithms and rules. Instead of manually creating each element, procedural modeling relies on mathematical functions, parameters, and logical operations to define the geometry and appearance of objects. This approach offers numerous advantages, including scalability, flexibility, and the ability to create variations easily. Procedural models can be modified parametrically, enabling quick adjustments without redoing the entire design. As a result, procedural modeling is widely used in various industries, including video games, visual effects, architectural visualization, and simulation. Here, we use Blender's Geometry Nodes feature (Fig. 4) which enables procedural modeling by connecting nodes that manipulate input geometry, perform operations like transformations and deformations, and generate or modify mesh topology. Attribute and math nodes manage data and perform calculations, while input nodes provide user-defined parameters. Geometry nodes are used to procedurally generate both a colon model, and to distribute polyps across the surface of the colon. Starting with a single curve to represent the shape of the colon, the entire model, including the location and size of polyps, is generated from scratch (Fig. 5).

2.3 Rendering/Synthetic Dataset Generation

For each frame of data, texture parameters were randomised controlling displacement magnitude, bump map magnitude, colour of the organs/tissues, subsurface scattering parameters, noise levels etc. The intensity of the lighting, the level of motion blur, and the position, look direction and focal distance of the camera

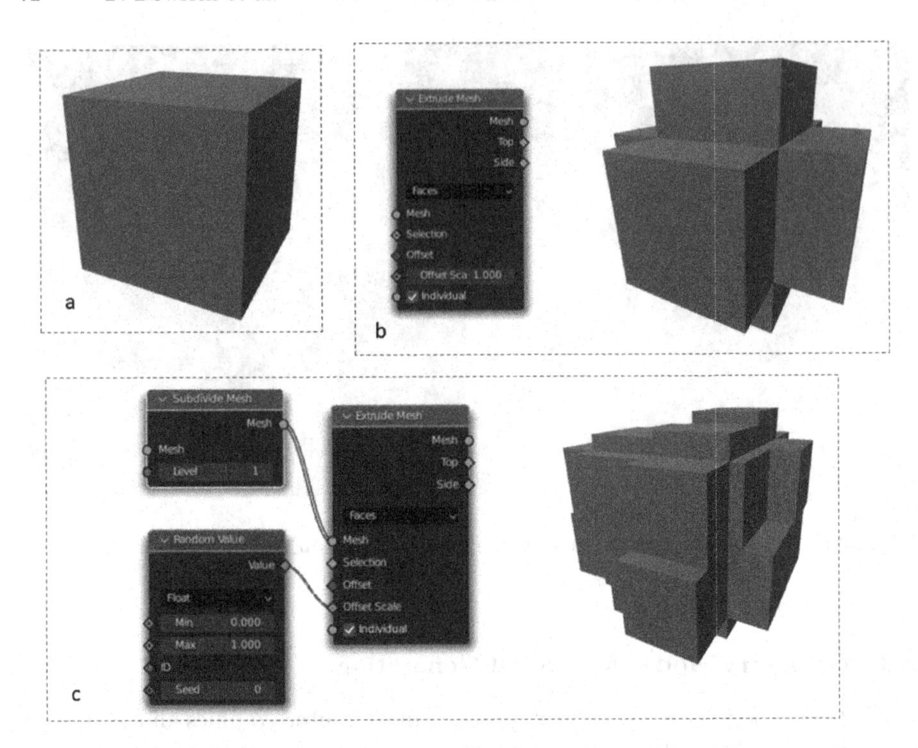

Fig. 4. Geometry Nodes example. (a) Input Cube. (b) Linear extrusion of each face. (c) Subdivision of the mesh, then extrusion of each face to a random distance. Output geometry data is generated on the fly from as the input geometry/parameters are changed.

were also adjusted. All images were rendered using the Cycle raytracing engine, at dimensions of 512×512, with a noise threshold of 0.1 and 256 samples. Render time for each frame was $\sim 5\,\mathrm{s}$.

Colonoscopy - 50,000 frames. Geometry Nodes parameters, controlling the shape of the colon, and the position, size and distribution of polyps were randomized per frame.

Laparoscopic liver surgery - 50,000 frames. Publicly available models for liver, gallbladder, etc. were used, and mesh primitives were used to represent other organs/tissues where appropriate (abdominal cavity = sphere etc.).

Custom Labelling. Semantic segmentations (Liver) were acquired by re-rendering the scene with each object assigned a flat colour, with bounding boxes (Polyps) derived from the minimum/maximum extents of the segmentation information.

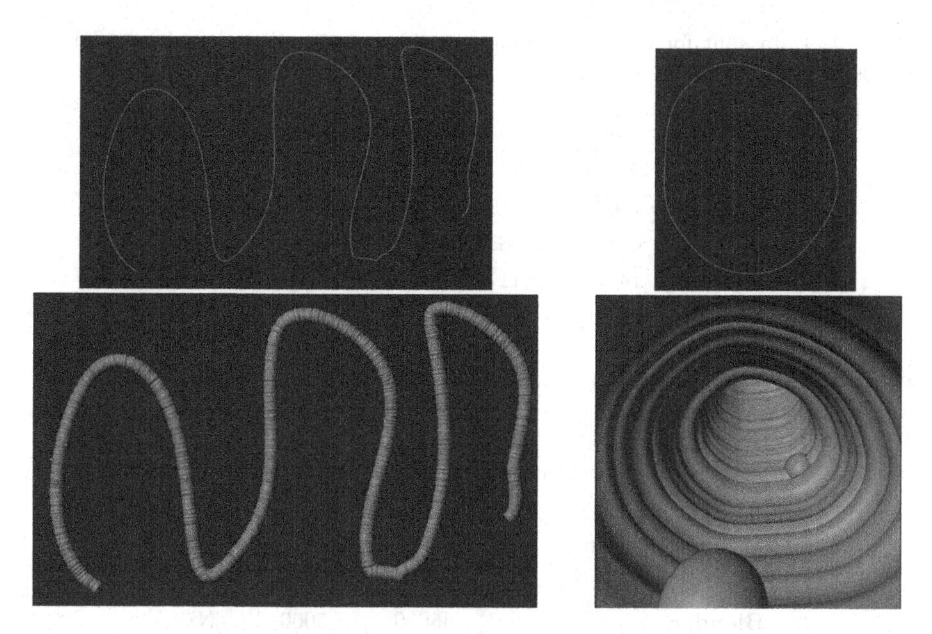

Fig. 5. Procedural generation of colon model. Clockwise from top left - curve used to define shape of colon; curve used to define cross section of colon; internal view of generated colon (no shading) with randomly placed polyps; external view of generated colon. The colon shape will update in real time in response to changes to either of the input curves. Further customization is carried out through randomization of Geometry Node parameters at simulation time.

2.4 Evaluation Datasets

Colonoscopy. Four datasets were used: Kansas Polyp Dataset [7], HyperKvasir [2], LDPolypVideo [8] and PolypGen [1]. Only polyp detection was considered, so from each data set, the relevant subset of data was used (HyperKvasir for example also contains upper GI tract data). Labels were converted to the COCO format. Kansas and LD datasets have a train/test/validation split already defined, which was left unchanged. LD data was split 80/10/10, and PolypGen was randomly split into 5 patients for train, 1 for test and 1 for validation (Table 1).

Laparoscopic Liver Segmentation. Three datasets were used: Dresden Surgical Anatomy Dataset (DSAD) [3], CholecSeg8k [5], SISVSE [16]. DSAD contains both single organ labelling and multi-organ labelling datasets. For this work, only the data from the liver single organ subset was used. The full SISVSE and CholecSeg8k datasets were used, with any non-liver labels removed.

Each dataset provides data from a number of separate patients/procedures. Data was randomly split into training, test and validation data, with an approximate 80/10/10 split of images between the three (Table 2). Actual splits deviate slightly from this, as the number of images for each patient varies. It should be noted that while CholecSeg8k has the highest number of images, the dataset

Table 1. Data split for colonoscopy data. Brackets indicate the number of labeled polyps in that set.

	Total Images	Train	Validation	Test
Kansas	37899	28773 (27048)	4254 (4214)	4872 (4719)
HyperKvasir	1000	800 (972)	100 (121)	100 (113)
LD	4186	20855 (18900)	3934 (4569)	15397 (15268)
PolypGen	1471	1178 (1191)	88 (98)	208 (204)
Blender	50000	45000 (60827)	5000 (7685)	N/A

Table 2. Data split for laparoscopy data. Brackets indicate the number of distinct patients/procedures in that set.

	Total Images	Train	Validation	Test
DSAD	1430 (23)	1131 (18)	101 (2)	119 (3)
CholecSeg8k	8080 (19)	6080 (15)	1000 (2)	2000 (2)
SISVSE	4510 (40)	3588 (32)	457 (4)	462 (4)
Blender	50000	45000	5000	N/A

consists of multiple sets of sequential frames taken from the same procedures, whereas each frame in DSAD and SISVSE are non-sequential/from different procedures.

2.5 Model Training

Laparoscopy - semantic segmentation Semantic segmentation was evaluated using a standard UNet configuration, with combined DICE loss and Cross Entropy loss, RMSprop optimizer, and learning rate of 1e–5. The network was trained on each dataset individually, as well as with pre training on Blender data for each dataset. Pre-training on Blender data was for 10 epochs; all other training runs were 50 epochs. This resulted in 6 trained models (each dataset with and without Blender pre-training), each of which was evaluated on the three sets of test data, with the DICE score for liver classification recorded.

Colonoscopy - polyp bounding box detection Polyp detection was trained using Yolov7 [12]. Default training parameters were used for the full YoloV7 network, with pre-trained ImageNet weights loaded. A model was trained for 100 epochs on each of the 4 datasets, as well as being pre-trained on the Blender data and post trained on each dataset. This resulted in 8 trained models (each dataset with and without Blender pre-training), each of which was evaluated on the four sets of test data.

The metrics reported by YoloV7 are precision, recall and mean average precision (mAP). mAP is calculated both for a single IoU of 0.5 (mAP@.5), and as an average of the mAP for IoU values between 0.5 and 0.95 (mAP@[.5:.95]).

Table 3. Liver segmentation DICE score, out of 100. Rows indicate training dataset, columns the test dataset. Cells with highlighted background show the highest value for that metric, across all models. Bold values indicate the highest value when the dataset used for training is excluded (e.g. excluding Cholec trained models from evaluation on Cholec data)

	Cholec	DSAD	SISVSE
Cholec	75	74	73
DSAD	37	85	61
SISVSE	23	77	77
Blender + Cholec	79	87	78
Blender + DSAD	55	96	**79**
Blender + SISVSE	**71**	**92**	91

3 Results

For laparoscopy data (Table 3), the use of synthetic data for pre-training increased the DICE score in 8 out of 9 cases, with the average change being an increase of 15% (p=0.02, using paired t-test).

For colonoscopy data (Table 4), for each evaluation metric, for each dataset, the highest value was achieved when the synthetic data was used for pre-training (cells with shaded background). If the training dataset is excluded from evaluation, 11 out of 16 metrics are achieved on pre-trained data (bold text in table); 3 are unchanged, and 2 are lower following pre-training. When the performance of individual metrics is compared with/without pre-training, then Precision is increased 12/16 times (Average change +11%, p=0.01), Recall 14/16 (+9%, p=0.002), MAP@.5 13/16 (+10%, p=0.01), MAP@[.5:.95] 13/16 (+8%, p=0.003).

4 Discussion

For both the laparoscopy (Table 3) and colonoscopy (Table 4) datasets, the use of synthetic data improved model performance, across all metrics, compared with train-ing only on real data. The results given in this work show that the method employed for procedural generation of training data can be used to improve model performance. It is envisaged that such methods would be complementary to existing approaches for data synthesis (GANs, diffusion models etc.) either by the use of multiple sources of synthetic data for training, or for example, by generating target geometries and labels using Geometry Nodes, and then applying an alternative method for texture synthesis. Being able to generate synthetic data in this way also extends the use of synthetic data to areas where there may not be sufficiently large training datasets to utilize deep learning methods. Further work is underway to consider the effects of changing the ratio

Table 4. Polyp detection results. All values given as a score out of 100. Columns represent the results on the test sets, and rows are the different trained models. Cells with a highlighted background show the highest value for that metric, across all models. Bold values indicate the highest value when the dataset used for training is excluded (e.g. excluding HyperKvasir models from evaluation on HyperKvasir test set). B = Blender, KA = Kansas, KV = Kvasir, L = LD, P = PolypGen.

	Kansas (KA)				Kvasir (KV)				LD (L)				PolypGen (P)			
	P	R	mAP .5	mAP [.5:.95]	P	R	mAP .5	mAP [.5:.95]	P	R	mAP .5	mAP [.5:.95]	P	R	mAP .5	mAP [.5:.95]
KA	83	74	82	49	**86**	54	61	40	55	36	37	17	45	33	33	16
KV	46	23	23	12	82	66	73	44	37	15	14	07	20	28	16	08
LD	64	38	41	24	84	67	74	46	**69**	47	52	24	54	43	44	21
P	**81**	39	50	**31**	**86**	74	80	55	52	34	36	18	69	50	59	37
B+KA	88	83	92	58	80	62	66	44	62	**42**	**45**	**21**	59	45	43	24
B+KV	65	41	44	28	93	80	84	64	60	38	41	20	**74**	**52**	**63**	**39**
B+LD	68	29	33	20	83	65	73	44	73	48	55	26	59	44	45	25
B+P	74	**43**	**50**	30	**86**	**81**	**83**	**62**	63	41	**45**	**21**	71	66	67	45

of synthetic to real data when training, and to make use of Geometry Nodes to provide more fine-grained labels, such as polyp sizing, for more advanced applications.

Acknowledgments. This work was funded by EPSRC (EP/V052438/1), and supported by a Microsoft Azure Research Grant, and an Oracle Cloud Computing Grant.

References

1. Ali, S., et al.: Polypgen: a multi-center polyp detection and segmentation dataset for generalisability assessment (June 2021)
2. Borgli, H., et al.: Hyperkvasir, a comprehensive multi-class image and video dataset for gastrointestinal endoscopy. Sci. Data **7** (2020). https://doi.org/10.1038/s41597-020-00622-y
3. Carstens, M., et al.: The dresden surgical anatomy dataset for abdominal organ segmentation in surgical data science. Sci. Data **10**, 3 (2023). https://doi.org/10.1038/s41597-022-01719-2
4. Funke, I., et al.: Generating large labeled data sets for laparoscopic image processing tasks using unpaired image-to-image translation. CoRR (2019)
5. Hong, W.Y., Kao, C.L., Kuo, Y.H., Wang, J.R., Chang, W.L., Shih, C.S.: Cholecseg8k: a semantic segmentation dataset for laparoscopic cholecystectomy based on cholec80 (Nov 2020)
6. Jagtap, A.D., Heinrich, M., Himstedt, M.: Automatic generation of synthetic colonoscopy videos for domain randomization (May 2022)
7. Li, K., et al.: Colonoscopy polyp detection and classification: dataset creation and comparative evaluations. PLoS ONE **16**, e0255809 (2021). https://doi.org/10.1371/journal.pone.0255809

8. Ma, Y., Chen, X., Cheng, K., Li, Y., Sun, B.: LDPolypVideo Benchmark: A Large-Scale Colonoscopy Video Dataset of Diverse Polyps, pp. 387–396 (2021). https://doi.org/10.1007/978-3-030-87240-3_37
9. Moreu, E., McGuinness, K., O'Connor, N.E.: Synthetic data for unsupervised polyp segmentation (Feb 2022)
10. Hinterstoisser, S., Pauly, O., Heibel, H., Marek, M., Bokeloh, M.: An annotation saved is an annotation earned: using fully synthetic training for object instance detection. In: Computer Vision and Pattern Recognition (2019)
11. Rivoir, D., et al.: Long-term temporally consistent unpaired video translation from simulated surgical 3d data (2021). https://doi.org/10.1109/ICCV48922.2021.00333
12. Wang, C.Y., Bochkovskiy, A., Liao, H.Y.M.: Yolov7: trainable bag-of-freebies sets new state-of-the-art for real-time object detectors (July 2022)
13. Wood, E., Baltrušaitis, T., Hewitt, C., Dziadzio, S., Cashman, T.J., Shotton, J.: Fake it till you make it: face analysis in the wild using synthetic data alone (2021). https://doi.org/10.1109/ICCV48922.2021.00366
14. Yonghao Long, Siu Hin Fan, Q.D.Y.W.: Neural rendering for stereo 3d reconstruction of deformable tissues in robotic surgery
15. Yoon, D., et al.: Colonoscopic image synthesis with generative adversarial network for enhanced detection of sessile serrated lesions using convolutional neural network. Sci. Rep. **12**, 261 (2022). https://doi.org/10.1038/s41598-021-04247-y
16. Yoon, J., et al.: Surgical Scene Segmentation Using Semantic Image Synthesis with Virtual Surgery Environment, pp. 551–561 (2022). https://doi.org/10.1007/978-3-031-16449-1_53

Improving Medical Image Classification in Noisy Labels Using only Self-supervised Pretraining

Bidur Khanal[1(✉)], Binod Bhattarai[4], Bishesh Khanal[3], and Cristian A. Linte[1,2]

[1] Center for Imaging Science, RIT, Rochester, NY, USA
bk9618@rit.edu
[2] Biomedical Engineering, RIT, Rochester, NY, USA
[3] NepAl Applied Mathematics and Informatics Institute for Research (NAAMII), Patan, Nepal
[4] University of Aberdeen, Aberdeen, UK

Abstract. Noisy labels hurt deep learning-based supervised image classification performance as the models may overfit the noise and learn corrupted feature extractors. For natural image classification training with noisy labeled data, model initialization with contrastive self-supervised pretrained weights has shown to reduce feature corruption and improve classification performance. However, no works have explored: i) how other self-supervised approaches, such as pretext task-based pretraining, impact the learning with noisy label, and ii) any self-supervised pretraining methods alone for medical images in noisy label settings. Medical images often feature smaller datasets and subtle inter-class variations, requiring human expertise to ensure correct classification. Thus, it is not clear if the methods improving learning with noisy labels in natural image datasets such as CIFAR would also help with medical images. In this work, we explore contrastive and pretext task-based self-supervised pretraining to initialize the weights of a deep learning classification model for two medical datasets with self-induced noisy labels—*NCT-CRC-HE-100K* tissue histological images and *COVID-QU-Ex* chest X-ray images. Our results show that models initialized with pretrained weights obtained from self-supervised learning can effectively learn better features and improve robustness against noisy labels.

Keywords: medical image classification · label noise · learning with noisy labels · self-supervised pretraining · warm-up obstacle · feature extraction

1 Introduction

Medical image classification using supervised learning relies on large amounts of representative data with accurately annotated labels to achieve good gener-

Supplementary Information The online version contains supplementary material available at https://doi.org/10.1007/978-3-031-44992-5_8.

B. Bhattarai et al. (Eds.): DEMI 2023, LNCS 14314, pp. 78–90, 2023.
https://doi.org/10.1007/978-3-031-44992-5_8

alization. However, recent practices of crowd-sourcing for data labeling or automatically generating labels from patients' medical reports using algorithms, and the high variability among expert annotators introduce higher levels of label noise in medical datasets. Moreover, supervised deep learning is highly susceptible to label noise as the models can easily overfit the noisy labels, leading to corrupt representation learning and compromising generalizability [18,19,24,39]. Correcting label noise in large medical image datasets is expensive and requires extensive human resources and time-consuming protocols. Several methods for learning with noisy labels (LNL) have been introduced in natural image datasets to minimize the influence of label noise on the training [2,8,12,25,32,34]. Similar methods, with adjustments, have also been applied to medical image classification [15,26,36,44].

Many LNL methods rely on a warm-up phase, a small number of initial epochs during which the model is trained directly using all the noisy training data [8,15,25,27]. While the warm-up phase is important to kickstart the model and learn basic features important for proper separation of noisy labels from clean labels at a later phase [41,42], the high noise rate makes it challenging to avoid memorizing wrong labels and learning poor feature extractors. Zheltonozhskii et al. [42] referred to this issue as the "warm-up obstacle". One may use supervised pretraining to learn good feature extractors and train with noisy labels to mitigate the warm-up obstacle. However, this approach presents challenges in medical datasets due to the limited availability of large labeled datasets that closely align with the given new dataset. Alternatively, if existing medical datasets already contain valuable metadata such as gender and age information, one may pretrain to predict such auxiliary information before proceeding with training on the main task involving noisy labels. Such an approach could minimize feature corruption, as the auxiliary tasks are relatively straightforward and less likely to contain label noise. However, if the datasets lack such metadata, another approach is to use self-supervised learning techniques for pretraining to learn feature extractors, without relying on any labels.

Some studies have demonstrated the benefits of contrastive learning-based self-supervised pretraining to improve robustness against noisy labels in natural image datasets [41,42]. However, no extensive study has been conducted to investigate which self-supervised pretraining is suitable for a specific scenario, therefore providing no such prior knowledge that can be adapted and used in medical image classification. Additionally, medical images come with some caveats that make it challenging to apply various self-supervised techniques in medical datasets (discussed in Sect. 2.2).

In this work, we investigate contrastive learning and propose simple and intuitive pretext task-based self-supervised pretraining approaches to improve robustness against noisy labels in the medical image classification problem. We show that self-supervised pretraining can significantly improve the robustness against noisy labels in the existing classification framework. Furthermore, we explored the implications of this pretraining approach on existing LNL methods by pretraining the models before the warm-up phase.

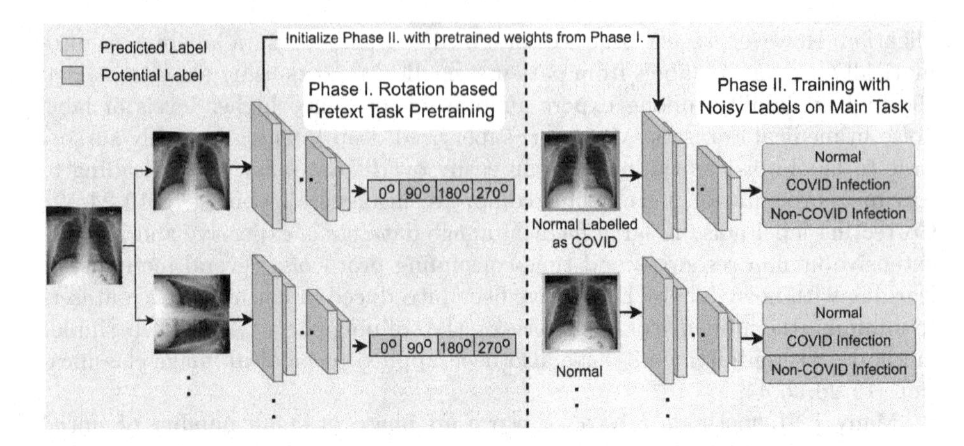

Fig. 1. Our approach involves two phases: I. Pretrain the model with pretext task-based self-supervised technique (left), and II. Retrain the pretrained model on medical image classification with noisy labels using LNL approaches (right).

Our contributions can be summarized as follows: **1)** To our knowledge, we are the first to investigate the use of only self-supervised pretraining to improve robustness in the presence of noisy labels for medical image classification; **2)** We propose the use of pretext task-based self-supervised pretraining in classification with noisy labels, which hasn't been studied even with natural image datasets; **3)** Using two representative datasets, namely X-ray and histopathology images, induced with label noise at various rates, we show that self-supervised pretraining alone improves the feature extractor, thus helping overcome the warm-up obstacle in LNL methods, yielding significantly improved performance while reducing the label memorization.

2 Related Works

2.1 Learning with Noisy Labels in Medical Images

Several methods have been proposed to robustly train medical image classifiers with noisy labels [16]. Pham et al. [30] used label smoothing to reduce the impact of noisy labels in thoracic disease classification. Dgani et al. [4] introduced a noise layer and modified the network architecture to address unreliable labels in breast classification. Le et al. [22] used a sample reweighting technique to robustly train a pancreatic cancer detection model with noisy labels, while Xue et al. [35] used a similar reweighting technique for skin lesion classification with noisy datasets.

Ju et al. [15] used dual-uncertainty estimation to tackle two cases: label noise due to disagreement among experts and single-target label noise, in skin lesions, prostate cancer, and retinal disease. Ying et al. [37] improved COVID-19 chest X-ray classification through techniques like PCA, low-rank representation, neighborhood graph regularization, and k-nearest neighbor. Similarly, Zhou et al.

[44] employed consistency regularization and disentangled distribution learning for multi-label disease classification and severity grading in chest X-rays and diabetic retinopathy. Xue et al. [36] combined a student-teacher network with a co-training strategy to improve prostrate cancer grading, skin classification, chest X-ray classification, and histopathology cancer detection, in different label noise settings. Liu et al. [26] proposed co-correcting, a curriculum learning-based label correction strategy, for robust training with noisy labels in metastatic tissue classification and melanoma classification.

Despite incorporating some concepts from self-supervised learning, no research has explored the impact of just self-supervised pretraining on enhancing robustness against noisy labels.

2.2 Self-supervised Pretraining

Several self-supervised techniques have emerged recently [7], encompassing simple pretext task-solving approaches [5,6,40], contrastive learning methods [3,10,38], and generative approaches [9,28,43]. Generative approaches have shown promise but face challenges due to training instability and high computational resource requirements. Additionally, the majority of recent generative approaches necessitate a Transformer as the backbone, and investigating the robustness of a Transformer-based architecture against label noise, in comparison to a CNN, is a distinct topic of discussion. Furthermore, recent mask image modeling-based generative approaches that learn by randomly masking a certain portion of the image may inadvertently miss crucial features. This issue could be problematic for medical datasets that rely on subtle image cues [13] and requires a separate investigation. Therefore, for this work, we considered focusing solely on contrastive learning, which is widely used, and the pretext task-based approach, which is simple but unexplored, leaving generative approaches for future investigation.

In this study, we chose three pretext tasks: *Rotation prediction*, *Jigsaw puzzle*, and *Jigmag puzzle*, and a contrastive approach: *SimCLR*. *Rotation prediction* [6] trains a model to predict the rotation degree of an image in various orientations. *Jigsaw puzzle* [29] requires training a model to learn to predict the arrangement of shuffled, non-overlapping patches in an image. *Jigmag puzzle* [20], originally proposed for histopathology images, learns to predict the arrangement of patches obtained from magnifying an image at various factors. *SimCLR* utilizes a contrastive loss to compare the representations of different augmented views of the same input, aiming to bring closer the augmented views (positive pairs) of the same image while keeping the augmented views of other images (negative pairs) far apart in the representation space. The benefit of pretraining depends on the suitability of the pretraining task with the main task [23].

We selected these pretext tasks because *Rotation prediction* and *Jigsaw puzzle* are commonly used in the literature, while *Jigmag puzzle* was specifically proposed to address the subtlety of medical images. Other pretext tasks, such as *Colorization* [21], were deemed unsuitable for our grayscale X-ray image and stained histopathology image datasets. *SimCLR* was chosen for the contrastive approach due to its simplified framework, which eliminates the need for special memory buffers or specialized architectures.

3 Datasets

3.1 COVID-QU-Ex

This dataset is a collection of chest X-ray images obtained from various patients [33] categorized into three groups: COVID infection, Non-COVID infection, and Normal. The dataset consists of a total of 27,132 training images, with 8,561 classified as Normal, 9,010 as Non-COVID-19, and 9,561 as COVID-19 cases. Additionally, there is an exclusive test set containing 6,788 images for evaluation.

3.2 NCT-CRC-HE-100K

This dataset has 100,000 histopathological image patches of size 224 × 224 extracted from stained tissue slides [17] featuring nine classes, such as adipose, lymphocytes, mucus, etc. The test set uses a different CRC-VAL-HE-7K dataset consisting of 7,180 images, featuring all nine classes of the training set.

4 Experimental Setup

4.1 Random Label Noise

To evaluate how a deep learning classifier performs on high label noise, we randomly flipped all the labels in the training set such that the original labels are assigned to any other labels within the close-set with some probability [14]. Assuming a training dataset $\{(\mathbf{x}_i, y_i)\}_i^n \in \mathcal{D}$, which contains n samples, x_i is a data point belonging to the set $\mathcal{X} \in \mathbb{R}^d$, and y_i is its corresponding class label from a close-set classes $C = \{c_0, c_1, .., c_4\}$. For any sample $(\mathbf{x}_i, y_i)$, we change its label y_i to $\hat{y}_i \overset{p}{\sim} C \setminus y_i$, where p is the noise probability and $C \setminus y_i$ denotes any label of close-set classes other than the true label. The label noise is symmetrical for all the classes within the close set. We conducted experiments using four different noise rates $p \in \{0.5, 0.6, 0.7, 0.8\}$.

As depicted in Fig. 2, the impact of noisy labels on test performance varies across datasets; *NCT-CRC-HE-100K* remains robust to noise below 0.5, whereas *COVID-QU-Ex* is affected at lower rates also.

4.2 Methodology

Our approach involves two stages: i) pretrain a model using self-supervised learning on the given dataset to learn meaningful feature extractors, and ii) train the pretrained model for medical image classification on the same dataset with noisy labels (Fig. 1). We primarily focus on the first stage, experimenting with four self-supervised tasks. In the second stage, we experimented with cross-entropy alone,

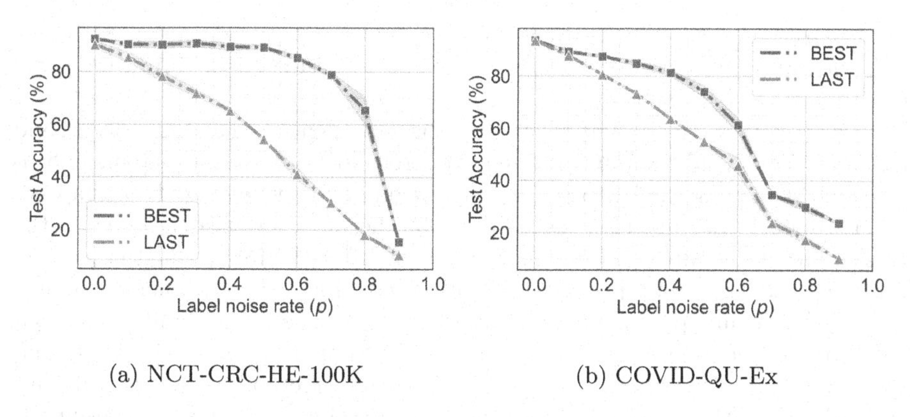

Fig. 2. Test performance as a function of training label noise rate (noise probability p) ranging from scale 0 to 1, with the shaded region indicating variability across three experimental trials. BEST indicates the highest test accuracy achieved, while LAST denotes the average test accuracy achieved in the last five epochs.

followed by two state-of-the-art LNL approaches, Co-teaching [8] and DivideMix [25].

Co-teaching selectively samples clean examples by ranking the training loss, while DivideMix utilizes a Gaussian Mixture Model (GMM) to categorize examples into clean and noisy groups based on the training loss of each sample. DivideMix also applies the MixMatch [1] semi-supervised learning approach by treating noisy labels as unlabeled examples. Notably, Co-teaching focuses on clean sample selection, while DivideMix does both clean sample selection and noisy label correction, but both use dual networks and utilize a warm-up phase.

Evaluation: Following [8,25], we evaluate the best test classification accuracy (BEST) and the average test accuracy of the last five epochs (LAST). The test set serves as a pseudo-test set, accessing the model's maximum performance with BEST, while LAST measures if the model has overfitted to noisy labels (see Fig. 2).

4.3 Implementation Details

Self-supervised Pretraining: We utilized the ResNet18 architecture for all experiments. For Rotation prediction, images were resized and underwent strong data augmentations: random horizontal flips, small rotations (10°), sharpness adjustment, equalization, and auto contrast. The model had to predict the rotation angle from four possible angles $(0°, 90°, 180°, 270°)$.

For Jigsaw puzzle solving, we performed similar strong augmentations and divided the resized input image into a 3×3 grid of patches. The patches were resized to 64×64 pixels, normalized with patch mean and standard deviation, and randomly shuffled to create one of the 1000 chosen permutations[1]. Then, they

[1] https://github.com/bbrattoli/JigsawPuzzlePytorch.

were passed through the ResNet18 feature extractor and concatenated before being fed into a fully connected output layer that predicted the input permutation.

For Jigmag puzzle solving, we applied the aforementioned augmentations and randomly magnified the input image at different locations using nine magnification factors ranging from 1 to 5. The magnified patches were resized, normalized, and randomly rearranged into one of the 1000 chosen permutations similar to the Jigsaw puzzle. A fully connected softmax layer after the ResNet feature extractor predicted the input permutation. We used SimCLR [3] implemented in[2], setting the default parameters. The input images were resized and augmented with random horizontal flips, color jitter, and Gaussian blur. Table 1 summarizes all training hyperparameter settings. All the methods were trained until convergence, with the number of epochs and learning rates chosen accordingly. For instance, Rotation prediction performed best with an SGD learning rate of 0.01 compared to other settings, while Jigsaw and Jigmag converged effectively using a batch size of 128.

Table 1. Hyperparameters used for training various self-supervised methods.

Datasets	Method	Input size	Batch	Epochs	Wt decay	Lr	Optim	Sheduler
NCT-CRC-HE-100K	Rotation	224 × 224	256	70	10^{-4}	0.01	SGD	Cosine Annealing
	Jigsaw	64 × 64	128	50	10^{-4}	0.001	Adam	Cosine Annealing
	Jigmag	64 × 64	128	50	10^{-4}	0.001	Adam	Cosine Annealing
	SimCLR	224 × 224	256	100	10^{-4}	0.001	Adam	Cosine Annealing
COVID-QU-Ex	Rotation	224 × 224	256	70	10^{-4}	0.01	SGD	Cosine Annealing
	Jigsaw	64 × 64	128	60	10^{-4}	0.001	Adam	Cosine Annealing
	Jigmag	64 × 64	128	60	10^{-4}	0.001	Adam	Cosine Annealing
	SimCLR	224 × 224	256	200	10^{-4}	0.001	Adam	Cosine Annealing

Learning with Noisy Labels: In this stage, we took the ResNet18 feature extractor initialized with from self-supervised training, added a fully connected output layer, and retrained the entire model with noisy labels. In the first set of experiments, we trained the model using standard cross-entropy loss without any modifications. The training process involved a batch size of 256, an SGD optimizer with a momentum of 0.9, weight decay of 10^{-4}, an initial learning rate of 0.01, and 50 training epochs. In the second set of experiments, we used two LNL methods.

[2] https://github.com/sthalles/SimCLR.

For Co-teaching, we followed the original paper's [8] recommendations and set the warm-up epochs to 10, $\tau = p$ and $c = 1$, where p is the label noise rate in data. As for DivideMix, we slightly adjusted the original hyperparameters [25], setting the warm-up epochs to 10, $M = 2$, $T = 0.2$, $\alpha = 4$, $\tau = 0.2$, and $\lambda_u = 0$ for $p = \{0.5, 0.6, 0.7\}$, while λ_u was changed to 0.25 for $p = 0.8$. Both methods maintained other training hyperparameters the same as the standard cross-entropy approach, except for DivideMix, where a batch size of 128 was used. To avoid confirmation bias, both Co-teaching and DivideMix original implementations initialize the dual networks with different weights. Similarly, in our approach, we adopt this strategy by initializing the dual networks with two distinct pretrained weights obtained from separate self-supervised training under the same settings.

All our experiments were implemented in Python 3.8 using the PyTorch 12.1.1 framework and trained on an A100 GPU (40 GB). We ran 3 experimental trials for each case to report the mean and standard deviation.

5 Results

Self-supervised Pretraining Improves Robustness Against Noisy Labels: In Fig. 3, we compared models trained with standard cross-entropy (CE) loss, using weights initialized from the self-supervised pretraining against PyTorch's default randomized He initialization [11].

The results demonstrate that self-supervised pretrained models significantly improve in terms of the BEST and LAST accuracy, particularly at high noise rates in the *NCT-CRC-HE-100K*. Specifically, SimCLR achieved better performance in both BEST and LAST accuracy at all noise rates, while Jigsaw and Jigmag also notably improve the LAST accuracy at $p = \{0.6, 0.7, 0.8\}$.

Similar trends are observed in the *COVID-QU-Ex*, where SimCLR, Jigmag, and Jigsaw outperform others significantly at $p = \{0.5, 0.6\}$ in terms of both BEST and LAST accuracy. At $p = \{0.7, 0.8\}$, rotation performs better, but the improvements are not as good as those observed with SimCLR, Jigsaw, and Jigmag in the $p = \{0.5, 0.6\}$ range. However, SimCLR, Jigmag, and Jigsaw performed worst than the cross entropy in the range $p = \{0.7, 0.8\}$.

The choice of the best self-supervised task varies based on the dataset, noise rate, and evaluation criteria, but the results achieved using self-supervised pretraining consistently show better performance compared to directly starting training from PyTorch's default randomized He initialization.

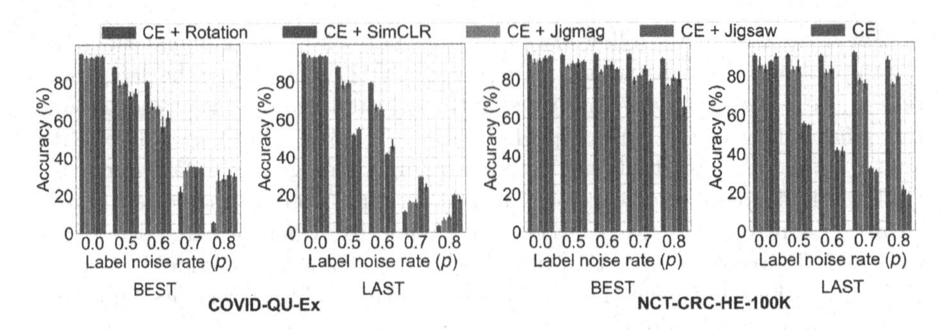

Fig. 3. Performance comparison of models trained starting from default weights vs. trained from weights initialized from self-supervised pretraining, when using standard cross entropy (CE) loss at different training label noise rates (p), in *COVID-QU-Ex* and *NCT-CRC-HE-100K*. BEST denotes the best test accuracy, while LAST denotes the average test accuracy achieved of the last five epochs. The experiments were run for three experimental trials to report the error bar.

Self-supervised Pretraining for LNL Methods: We compared LNL methods trained from PyTorch's default He initialization with those trained using weights initialized from a self-supervised pretraining in Fig. 4. The results show that the LNL methods already achieved good performance in the *NCT-CRC-HE-100K*, with only a small room for improvement at a lower noise rate. At higher noise rates, SimCLR achieved the highest BEST and LAST accuracy, followed by Rotation prediction. Initializing LNL with Jigsaw and Jigmag didn't improve the performance, but rather degraded it.

In the *COVID-QU-Ex*, we observed an improvement in classification performance in the noise range $p = \{0.5, 0.6\}$ for both Coteaching and DivideMix when using pretrained weights from the SimCLR and Rotation prediction task. However, beyond $p = \{0.7, 0.8\}$, the scores exhibited high variability, making it difficult to identify the best performance. SimCLR struggled with high noise rates, possibly due to excessive contrastive learning augmentations that overlooked vital subtle features valuable for discerning classes at high label noise. But, this speculation needs further study. Additionally, in *COVID-QU-Ex*, we noticed that the clean samples selected by LNL methods were biased towards one class and ignored the other classes, particularly at high noise levels, making it intriguing to investigate the cause behind this phenomenon in the future.

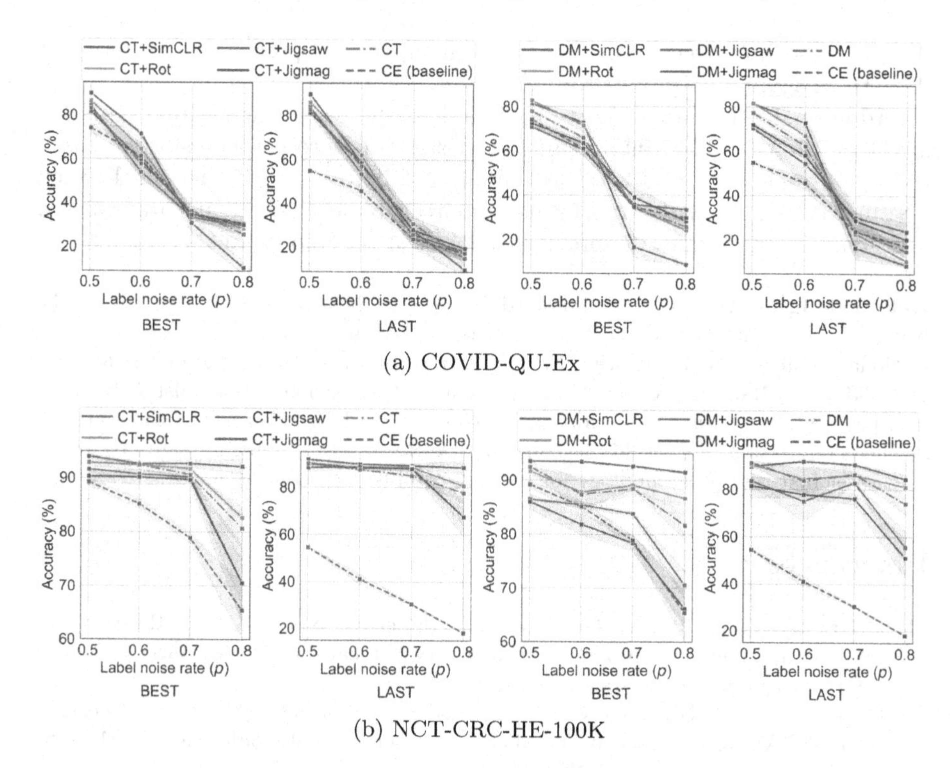

(a) COVID-QU-Ex

(b) NCT-CRC-HE-100K

Fig. 4. Performance comparison of existing LNL methods when initialized with self-supervised pretraining against baselines at different training label noise rates (p), in (a) *COVID-QU-Ex* and (b) *NCT-CRC-HE-100K*. Cross entropy (CE) denotes the actual baseline, Coteaching (CT) and Dividemix (DM) are existing LNL methods, and the term after + represents various self-supervised pretraining methods. BEST denotes the best test accuracy, while LAST denotes the average test accuracy achieved of the last five epochs. The shaded region along the line indicates the variability across three experimental trials.

6 Conclusion

We examined the effectiveness of utilizing self-supervised pretraining alone to improve the model's robustness against noisy labels in medical image classification. The choice of self-supervised task varied depending on the dataset, noise rate, and, evaluation criteria, with SimCLR consistently yielding the best results in most cases with LNL.

This study addresses a gap in the current research by serving as a first demonstration of the benefits of various self-supervised pretraining for medical image classification with noisy labels and offering valuable insights to mitigate the impact of high label noise. In the future, we plan to investigate additional self-supervised baselines and further explore how the nature and size of the dataset influence the improvements offered by self-supervised pretraining in terms of robustness against noisy labels.

Additionally, in this work, we have limited the investigation to CNN-based architecture. It would be interesting to investigate how recent Transformer-based architectures behave under various levels of label noise, whether off-the-shelf LNL methods work with Transformer architecture, and how Transformer-based self-supervised techniques improve robustness against noisy labels.

Acknowledgments. Research reported in this publication was supported by the National Institute of General Medical Sciences Award No. R35GM128877 of the National Institutes of Health, the Office of Advanced Cyber Infrastructure Award No. 1808530 of the National Science Foundation, and the Division Of Chemistry, Bioengineering, Environmental, and Transport Systems Award No. 2245152 of the National Science Foundation. We would like to thank the Research Computing team [31] at the Rochester Institute of Technology for proving computing resources for this research.

References

1. Berthelot, D., Carlini, N., Goodfellow, I., Papernot, N., Oliver, A., Raffel, C.A.: Mixmatch: a holistic approach to semi-supervised learning. In: Advances in Neural Information Processing Systems 32 (2019)
2. Chen, P., Liao, B.B., Chen, G., Zhang, S.: Understanding and utilizing deep neural networks trained with noisy labels. In: International Conference on Machine Learning, pp. 1062–1070. PMLR (2019)
3. Chen, T., Kornblith, S., Norouzi, M., Hinton, G.: A simple framework for contrastive learning of visual representations. In: International Conference on Machine Learning, pp. 1597–1607. PMLR (2020)
4. Dgani, Y., Greenspan, H., Goldberger, J.: Training a neural network based on unreliable human annotation of medical images. In: 2018 IEEE 15th International Symposium on Biomedical Imaging (ISBI 2018), pp. 39–42. IEEE (2018)
5. Doersch, C., Gupta, A., Efros, A.A.: Unsupervised visual representation learning by context prediction. In: Proceedings of the IEEE International Conference on Computer Vision, pp. 1422–1430 (2015)
6. Gidaris, S., Singh, P., Komodakis, N.: Unsupervised representation learning by predicting image rotations. arXiv preprint arXiv:1803.07728 (2018)
7. Gui, J., Chen, T., Cao, Q., Sun, Z., Luo, H., Tao, D.: A survey of self-supervised learning from multiple perspectives: Algorithms, theory, applications and future trends. arXiv preprint arXiv:2301.05712 (2023)
8. Han, B., et al.: Co-teaching: Robust training of deep neural networks with extremely noisy labels. In: Advances In Neural Information Processing Systems 31 (2018)
9. He, K., Chen, X., Xie, S., Li, Y., Dollár, P., Girshick, R.: Masked autoencoders are scalable vision learners. In: Proceedings of the IEEE/CVF Conference on Computer Vision and Pattern Recognition, pp. 16000–16009 (2022)
10. He, K., Fan, H., Wu, Y., Xie, S., Girshick, R.: Momentum contrast for unsupervised visual representation learning. In: Proceedings of the IEEE/CVF Conference on Computer Vision and Pattern Recognition, pp. 9729–9738 (2020)

11. He, K., Zhang, X., Ren, S., Sun, J.: Delving deep into rectifiers: surpassing human-level performance on imagenet classification. In: Proceedings of the IEEE International Conference on Computer Vision, pp. 1026–1034 (2015)
12. Hu, W., Li, Z., Yu, D.: Simple and effective regularization methods for training on noisily labeled data with generalization guarantee. arXiv preprint arXiv:1905.11368 (2019)
13. Huang, S.-C., Pareek, A., Jensen, M., Lungren, M.P., Yeung, S., Chaudhari, A.S.: Self-supervised learning for medical image classification: a systematic review and implementation guidelines. NPJ Digital Med. **6**(1), 74 (2023)
14. Jiang, L., Zhou, Z., Leung, T., Li, L.-J., Fei-Fei, L.: Mentornet: learning data-driven curriculum for very deep neural networks on corrupted labels. In: International Conference on Machine Learning, pp. 2304–2313. PMLR (2018)
15. Lie, J., et al.: Improving medical images classification with label noise using dual-uncertainty estimation. IEEE Trans. Med. Imaging **41**(6), 1533–1546 (2022)
16. Karimi, D., Dou, H., Warfield, K., Gholipour, A.: Deep learning with noisy labels: exploring techniques and remedies in medical image analysis. Medical Image Anal. **65**, 101759 (2020)
17. Kather, J.N., et al.: Predicting survival from colorectal cancer histology slides using deep learning: a retrospective multicenter study. PLoS Med. **16**(1), e1002730 (2019)
18. Khanal, B., Kamrul Hasan, S.M., Khanal, B., Linte, C.A.: Investigating the impact of class-dependent label noise in medical image classification. In: Medical Imaging 2023: Image Processing, vol. 12464, pp. 728–733. SPIE (2023)
19. Khanal, B., Kanan, C.: How does heterogeneous label noise impact generalization in neural nets? In: Bebis, G., et al. (eds.) ISVC 2021. LNCS, vol. 13018, pp. 229–241. Springer, Cham (2021). https://doi.org/10.1007/978-3-030-90436-4_18
20. Koohbanani, N.A., Unnikrishnan, B., Khurram, S.A., Krishnaswamy, P., Rajpoot, N.: Self-path: self-supervision for classification of pathology images with limited annotations. IEEE Trans. Med. Imaging **40**(10), 2845–2856 (2021)
21. Larsson, G., Maire, M., Shakhnarovich, G.: Colorization as a proxy task for visual understanding. In: Proceedings of the IEEE Conference on Computer Vision and Pattern Recognition, pp. 6874–6883 (2017)
22. Le, H., Samaras, D., Kurc, T., Gupta, R., Shroyer, K., Saltz, J.: Pancreatic cancer detection in whole slide images using noisy label annotations. In: Shen, D., et al. (eds.) MICCAI 2019. LNCS, vol. 11764, pp. 541–549. Springer, Cham (2019). https://doi.org/10.1007/978-3-030-32239-7_60
23. Lee, J.D., Lei, Q., Saunshi, N., Zhuo, J.: Provable self-supervised learning: Predicting what you already know helps. Adv. Neural. Inf. Process. Syst. **34**, 309–323 (2021)
24. Lee, K., Yun, S., Lee, K., Lee, H., Li, B., Shin, J.: Robust inference via generative classifiers for handling noisy labels. In: International Conference on Machine Learning, pp. 3763–3772. PMLR (2019)
25. Li, J., Socher, R., Hoi, S.C.H.: Dividemix: learning with noisy labels as semi-supervised learning. In: International Conference on Learning Representations (2020)
26. Liu, J., Li, R., Sun, C.: Co-correcting: noise-tolerant medical image classification via mutual label correction. IEEE Trans. Med. Imaging **40**(12), 3580–3592 (2021)
27. Liu, S., Niles-Weed, J., Razavian, N., Fernandez-Granda, C.: Early-learning regularization prevents memorization of noisy labels. Adv. Neural. Inf. Process. Syst. **33**, 20331–20342 (2020)

28. Liu, X., et al.: Self-supervised learning: generative or contrastive. IEEE Trans. Knowl. Data Eng. **35**(1), 857–876 (2021)
29. Noroozi, M., Favaro, P.: Unsupervised learning of visual representations by solving jigsaw puzzles. In: Leibe, B., Matas, J., Sebe, N., Welling, M. (eds.) ECCV 2016. LNCS, vol. 9910, pp. 69–84. Springer, Cham (2016). https://doi.org/10.1007/978-3-319-46466-4_5
30. Pham, H.H., Le, T.T., Tran, D.Q., Ngo, D.T., Nguyen, H.Q.: Interpreting chest x-rays via cnns that exploit hierarchical disease dependencies and uncertainty labels. Neurocomputing **437**, 186–194 (2021)
31. Rochester Institute of Technology. Research computing services (2022)
32. Song, H., Kim, M., Park, D., Lee, J.-G.: How does early stopping help generalization against label noise? arXiv preprint arXiv:1911.08059 (2019)
33. Tahir, A.M et al.: COVID-QU-Ex Dataset (2022)
34. Wei, H., Feng, L., Chen, X., An, B.: Combating noisy labels by agreement: a joint training method with co-regularization. In: Proceedings of the IEEE/CVF Conference on Computer Vision and Pattern Recognition (CVPR) (June 2020)
35. Xue, C., Dou, Q., Shi, X., Chen, H., Heng, P.-A.: Robust learning at noisy labeled medical images: Applied to skin lesion classification. In: 2019 IEEE 16th International Symposium on Biomedical Imaging (ISBI 2019), pp. 1280–1283. IEEE (2019)
36. Xue, C., Lequan, Yu., Chen, P., Dou, Q., Heng, P.-A.: Robust medical image classification from noisy labeled data with global and local representation guided co-training. IEEE Trans. Med. Imaging **41**(6), 1371–1382 (2022)
37. Ying, X., Liu, H., Huang, R.: Covid-19 chest x-ray image classification in the presence of noisy labels. Displays, 102370 (2023)
38. Zbontar, J., Jing, L., Misra, I., LeCun, Y., Deny, S.: Barlow twins: self-supervised learning via redundancy reduction. In: International Conference on Machine Learning, pp. 12310–12320. PMLR (2021)
39. Zhang, C., Bengio, S., Hardt, M., Recht, B., Vinyals, O.: Understanding deep learning (still) requires rethinking generalization. Commun. ACM **64**(3), 107–115 (2021)
40. Zhang, R., Isola, P., Efros, A.A.: Colorful image colorization. In: Leibe, B., Matas, J., Sebe, N., Welling, M. (eds.) ECCV 2016. LNCS, vol. 9907, pp. 649–666. Springer, Cham (2016). https://doi.org/10.1007/978-3-319-46487-9_40
41. Zhang, X.: Codim: learning with noisy labels via contrastive semi-supervised learning. arXiv preprint arXiv:2111.11652 (2021)
42. Zheltonozhskii, E., Baskin, C., Mendelson, A., Bronstein, A.M., Litany, O.: Contrast to divide: Self-supervised pre-training for learning with noisy labels. In: Proceedings of the IEEE/CVF Winter Conference on Applications of Computer Vision, pp. 1657–1667 (2022)
43. Zhou, J.: ibot: image bert pre-training with online tokenizer. arXiv preprint arXiv:2111.07832 (2021)
44. Zhou, Y., Huang, L., Zhou, T., Sun, H.: Combating medical noisy labels by disentangled distribution learning and consistency regularization. Futur. Gener. Comput. Syst. **141**, 567–576 (2023)

Investigating Transformer Encoding Techniques to Improve Data-Driven Volume-to-Surface Liver Registration for Image-Guided Navigation

Michael Young[1](✉), Zixin Yang[1], Richard Simon[2], and Cristian A. Linte[1,2]

[1] Chester F. Carlson Center for Imaging Science, Rochester Institute of Technology, Rochester, NY 14623, USA
{may1514,yy8898,calbme}@rit.edu

[2] Department of Biomedical Engineering, Rochester Institute of Technology, Rochester, NY 14623, USA
rasbme@rit.edu

Abstract. Due to limited direct organ visualization, minimally invasive interventions rely extensively on medical imaging and image guidance to ensure accurate surgical instrument navigation and target tissue manipulation. In the context of laparoscopic liver interventions, intra-operative video imaging only provides a limited field-of-view of the liver surface, with no information of any internal liver lesions identified during diagnosis using pre-procedural imaging. Hence, to enhance intra-procedural visualization and navigation, the registration of pre-procedural, diagnostic images and anatomical models featuring target tissues to be accessed or manipulated during surgery entails a sufficient accurate registration of the pre-procedural data into the intra-operative setting. Prior work has demonstrated the feasibility of neural network-based solutions for nonrigid volume-to-surface liver registration. However, view occlusion, lack of meaningful feature landmarks, and liver deformation between the pre- and intra-operative settings all contribute to the difficulty of this registration task. In this work, we leverage some of the state-of-the-art deep learning frameworks to implement and test various network architecture modifications toward improving the accuracy and robustness of volume-to-surface liver registration. Specifically, we focus on the adaptation of a transformer-based segmentation network for the task of better predicting the optimal displacement field for nonrigid registration. Our results suggest that one particular transformer-based network architecture—UTNet—led to significant improvements over baseline performance, yielding a mean displacement error on the order of 4 mm across a variety of datasets.

Keywords: Nonrigid Registration · Laparoscopy · Machine Learning · Transformer · Neural Network

B. Bhattarai et al. (Eds.): DEMI 2023, LNCS 14314, pp. 91–101, 2023.
https://doi.org/10.1007/978-3-031-44992-5_9

1 Introduction

Background and Motivation: Hepatocellular carcinoma is a pressing concern in oncology, being the fifth-most common cancer responsible for the second-most cancer-related deaths [10]. For these cases, surgery is frequently the standard of care [16].

For all minimally invasive intervention applications, accurate navigation to relevant tissues is paramount. In laparoscopic surgery, the procedure is performed under guidance provided by a camera inserted through a small incision. While this confers several benefits such as recovery time, additional difficulties are encountered in surgical navigation. Limited field-of-view (FOV) and the homogeneous appearance of the surface of organs can pose significant difficulty in locating relevant lesions [21].

This task can be facilitated by using 3D preoperative scans, generated from Computed Tomography (CT) or Magnetic Resonance Imaging (MRI). While this approach has some benefits, the use of pre-procedural scans adds a necessary preprocessing step to be performed during surgery: the registration of that data to the surgical view. This step has several challenges that need to be overcome. For rigid registration, the homogeneous intraoperative surface and varying noise characteristics make localizing a specific view on the intraoperative liver difficult [22]. The nature of the liver as a soft body introduces additional difficulties after rigid registration. Factors such as the interaction of surgical instruments with the organs, patient breathing, and insufflation of the abdominal cavity during surgery to increase the working volume lead to deformations that must be predicted and compensated for in order to achieve a sufficiently accurate and faithful pre- to intra-operative organ registration [26]. In addition, the opacity of most organs implies that the intraoperative view cannot be easily modeled as a closed shape of finite volume. Rather, the problem of registering the preoperative scan onto a limited intraoperative view is a problem of volume-to-surface registration. This task entails two major components: first, a correspondence must be found between the partial surface and the complete volume; second, both rigid and nonrigid registration must be performed to correct for deformations between the preoperative organ volume (from CT or MRI) and the reconstructed intraoperative partial organ surface. Both tasks have been the focus of substantial prior work, both in the clinical setting [5,12,14] and elsewhere [30]. Prior work has identified the potential advantages of image-guided navigation in concert with augmented reality visualization during minimally invasive liver surgery. While currently proposed methods could be highly useful to the surgeon, improvements in anatomical precision are necessary to increase the value of image guidance in the operating theater [1,4]. Therefore, a rapid volume-to-surface registration method would enable the surgeon to visualize in real time the intraoperative location of relevant lesions present inside the liver, and identified in pre-procedural scans, but not visible using intraoperative video, since located beneath the liver surface, in turn, allowing for more effective visualization and navigation to the target tissue during surgery.

Prior Work: Prior literature indicates the viability of predicting soft-body deformations given partial data. Several of these methods function by input of two meshes representing the preoperative and intraoperative geometriesand output the deformation field that warps the preoperative geometry to match the intraoperative geometry [18,19]. Sulewack *et al.* [26] have developed a physics-based shape matching method for this task. While this does achieve sub-millimeter registration accuracy, the need for manual statement of boundary conditions and inference time hamper practical usefulness. Other methods have demonstrated the viability of lower-dimensional representations of 3D objects. Small-scale neural networks trained on individual scenes have allowed for efficient volume encoding and generation of novel views [7,15,20].

Recent advances in computer vision and image processing have focused on the network structure known as the Transformer, first described in [27]. This deviates from the CNN architecture by creating representations of patch sequences, and using self-attention to extract more global information. This in turn allows transformers to extract more global information, contrasted with the limited influence range of a CNN. Prior work showed its effectiveness in image classification tasks [8] and in image segmentation [11,28].

Proposed Work: The proposed work leverages the prior work of [22] that yielded V2S-Net, a Convolutional Neural Network (CNN) to simultaneously establish surface correspondences and perform the nonrigid registration in one step. Their implementation employs a structure akin to a U-Net as in [24]. It uses voxelized representations of the preoperative volume and intraoperative surface as input, and generates a $3 \times 64 \times 64 \times 64$ voxel image corresponding to the spatial displacement components. Such an implementation allows for efficient inferencing and simple scalability for large quantities of synthetic data.

In this work, we build on the technique proposed by Pfeiffer *et al.* [22] by investigating several alterations to the network architecture to more accurately estimate the pre- to intraoperative displacement to help achieve a better registration. The most promising network architecture modification found, and the focus of this work, consists of the use of transformer architectures to better encode global shape information, which, in turn, will provide better control toward better predicting the pre- to intra-operative displacement field.

Following the example in [11], our proposed UTNet-inspired architecture is adapted for this 3D image transformation task by employing transformer encoder blocks on the encoding pathway and replacing the traditional skip connections with transformer decoding blocks. Further investigation consists of altering network components such as activation function and the presence of dropout. Finally, the performance of the proposed network architectures is evaluated by assessing their accuracy (and robustness) achieved under different levels of noise present in the test data.

2 Methods

Training Data: Training data for the networks in this study were generated using the pipeline in [22]. The pipeline begins by generating an icosphere in Blender [6] and uses automated operations to deform it into a soft body of random shape. Figure 1 shows one such random body. In our work, we modify the previous implementation at this step by repairing non-closed meshes; this enforcement of watertightness improves stability in generating valid data. Gmsh [13] is then used to convert the surface mesh to a tetrahedral volumetric mesh. Random forces of 1.5 N maximum magnitude are assigned to specific locations on the mesh surface, and zero-displacement boundary conditions are applied to randomly-selected areas. These data are passed to Elmer [17] to calculate the displacement field via the Finite Element Modeling (FEM) method.

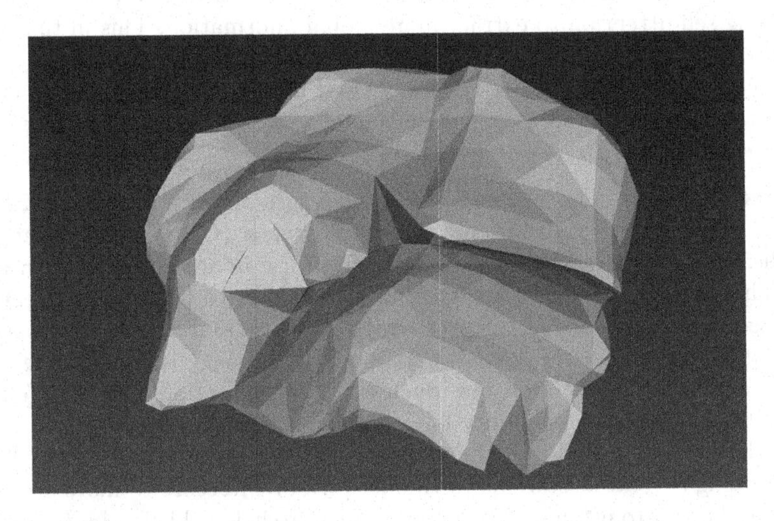

Fig. 1. An example of a random body surface generated by deforming the icosphere using the training dataset and pipeline proposed in [22]

The FEM yields the equivalent of an intraoperative organ volume used to extract a random surface point cloud patch to serve as an intraoperative limited laparoscopic view; in addition, random portions of the patch are removed to simulate occlusion. Lastly, to better portray the reality of intraoperative data, uncertainty is added to the dataset by displacing 30% of the surface points along each axis by uniform noise with a magnitude of no more than 1 cm.

In order to easily use this data as a neural network input, both the preoperative and intraoperative surfaces are voxelized. A uniform $64 \times 64 \times 64$ grid of 30 cm in each direction is generated for the preoperative and intraoperative surfaces. In each case, each voxel represents the shortest distance from the center of that voxel to the surface. For the preoperative case, the sign of the distance map is inverted for voxels inside the surface. The displacement field is voxelized in a similar manner using Gaussian interpolation. For the purpose of our work, a total of 40 000 cases are generated in this fashion. Further data augmentation consists of reflecting the samples across the xy, yz, and xz planes, scaling the training set size by a factor of 8 and yielding approximately 32 0000 effective training samples. Figure 2 shows the summary of this process. Dataset used is available upon request.

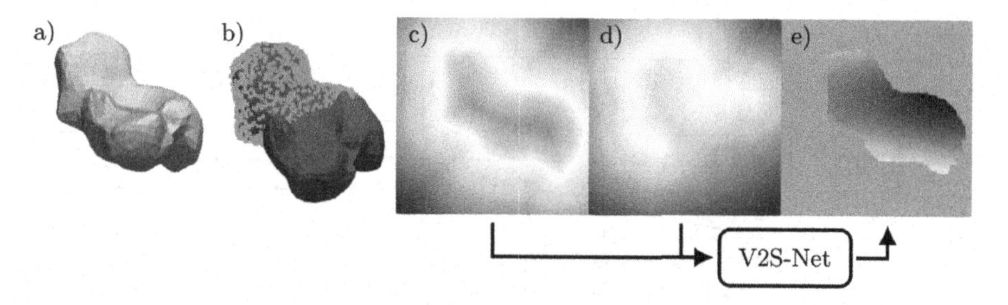

Fig. 2. A diagram of the data generation pipeline, reproduced from [22]. a) Preoperative volume mesh; b) Intraoperative surface in green with partial surface in orange; c) Preoperative signed distance map; d) Intraoperative distance map; e) Ground truth displacement field to be predicted.

Testing Data: In order to further evaluate the robustness of the network, a number of additional datasets are generated.

To evaluate network performance in the presence of additional noise, a dataset of 1000 samples was created by adding displacement noise featuring a maximum magnitude of 5 cm. An additional dataset was generated without noise to assess the ability of the network to encode clean, noiseless shapes.

In order to assess ability to generalize to liver shapes specifically, two additional datasets were generated based on previously generated liver meshes. One dataset is based on a set of 120 liver meshes derived from liver data in [2]. To augment the dataset, each liver mesh was scaled by 5 random scaling factors. The above pipeline was employed to generate a total of 1200 testing samples. A second liver dataset was derived from the liver samples used by Suwelack et al. in [26] in concert with a Physics-Based Shape Matching (PBSM) method. Mesh representations of the liver phantoms used therein were obtained and used to generate a series of four additional test cases.

Network Structures: For additional validation, the original V2S-Net network was re-run with the newly-generated training dataset. This network features a CNN architecture that uses an encoder chain to capture global detail, a decoder chain to return to output resolution and skip connections to carry over higher-resolution details to the decoder chain. Figure 3 shows a diagram of the network structure. Elementary changes to the network were investigated by generating two additional networks with similar structure: one using the Rectified Linear Unit (ReLU) activation function at non-output layers, and one including a dropout layer with probability 20% at each level of the encoder chain. Prior work has seen performance improvements with either change: see Sivagami *et al.* [25] and Yang *et al.* [29], respectively.

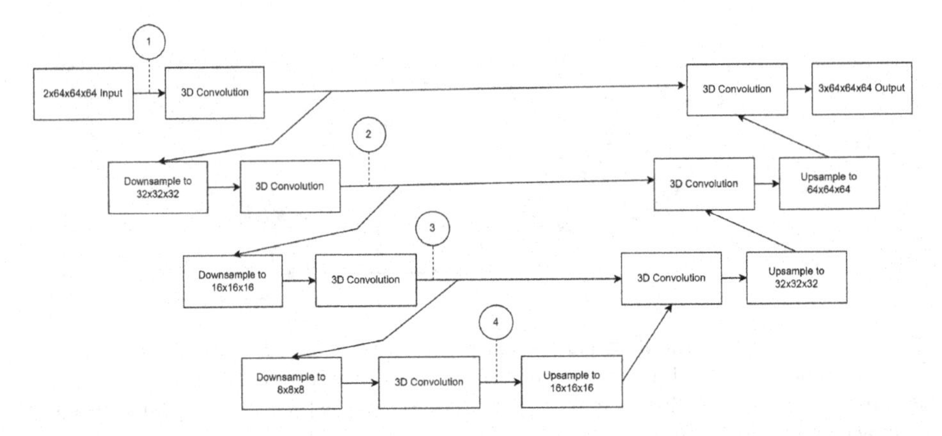

Fig. 3. A diagram of the general network structure of V2S-Net. The circled numbers indicate locations where relevant structures are appended to modify the original network (V2S-Net) to generate modified networks evauated in this work: the Input network, featuring a vision transformer at location 1; the Bottleneck network, featuring a vision transformer at location 4; and the ViT network, featuring a vision transformer at locations 1, 2, 3, and 4.

The original V2S-Net framework was further modified by the addition of the transformer module as shown in [8]. Input and output channels were chosen to maintain parity with the original network. The networks generated in this fashion are as follows: the *Input* network, with a vision transformer at location 1; the *Bottleneck* network, with a vision transformer at location 4; and the *ViT* network, with a vision transformer at locations 1, 2, 3, and 4.

An additional network, modeled after the UTNet framework in [11], was also constructed. This network uses a similar methodology as the *ViT* network, but alters the skip connections to instead employ a transformer decoder block to combine upsampled features with features from the encoder chain. In light of the prior work by Gao *et al.* [11], it is hypothesized that including transformer architectures within the network will allow for more efficient encoding of shape information similar to the semantic encoding described in [28]. This approach would, in turn, yield more efficient training and more accurate estimates without risk of overtraining due to the additional parameters.

Networks were trained using the research computing cluster at Rochester Institute of Technology [23]. A one-cycle learning rate scheduler and the Adam optimizer were used to train each network for 100 epochs.

Evaluation: We assessed the performance of all modified network architectures against the performance of the original network architecture (V2S-Net), in terms of the accuracy of their predicted displacement fields relative to the ground truth displacement field. Specifically, we computed the mean displacement error (MDE) in mm, as the difference between the displacement field predicted by each network architecture and the ground truth displacement field. The MDE was averaged across each testing set for each network architecture. In addition, to compare the performance of the modified network architectures to that of the original baseline (V2S-Net) network, we conducted statistical tests to identify any statistically significant differences in performance (quantified by the MDE metric) brought forth by the network modifications under investigation.

3 Results and Discussion

Table 1. Summary of Mean Displacement Error (MDE) in mm reported as mean $\pm$ standard error, computed between the predicted displacement and ground-truth displacement achieved by each model configuration under investigation and across all datasets used for training and validation

	Mean Displacement Error (MDE): Mean ± Std. Err. (mm)				
Model/Dataset	Synthetic Validation Set	Liver Test Set	PBSM Dataset	Noise Free Synthetic Data	High Noise Synthetic Data
V2S-Net	5.4 ± 0.5	4.02 ± 0.09	$\mathbf{2.9 \pm 0.6}$	5.4 ± 0.2	5.6 ± 0.3
Bottleneck	5.9 ± 0.6	4.16 ± 0.09	4.0 ± 0.8	5.6 ± 0.2	5.7 ± 0.2
Input	5.6 ± 0.5	4.10 ± 0.09	3.2 ± 0.7	5.4 ± 0.2	5.6 ± 0.3
ViT	5.2 ± 0.5	4.16 ± 0.09	3.2 ± 0.5	5.4 ± 0.2	5.6 ± 0.2
UTNet	$\mathbf{4.7 \pm 0.5}$	3.91 ± 0.07	3.9 ± 1.3	$\mathbf{4.9 \pm 0.2}$	$\mathbf{5.0 \pm 0.2}$
V2S-Net (ReLU)	15.8 ± 1.3	7.2 ± 0.1	6.0 ± 1.0	14.5 ± 0.4	14.5 ± 0.4
V2S-Net (dropout)	5.1 ± 0.5	$\mathbf{3.73 \pm 0.05}$	3.0 ± 0.5	5.1 ± 0.2	5.5 ± 0.3

In general, the implementation of the UTNet network yields lower MDE across the various testing datasets (see Table 1, Fig. 4). However, the high variability in MDE across all networks limits the conclusiveness of this difference. Pfeiffer *et al.* [22] noted that outliers could be observed during testing, especially for cases with relatively low visible surface area. Contrary to expectations, simply implementing the vision transformer modules do not appear to significantly improve MDE, and seemingly leads to slight degradation in some cases. In this case, it appears that under-generalization caused by the increased number of parameters outweighs the benefits of the transformer architecture. Nevertheless the UTNet-based architecture, with the most parameters of all, displays a generally lower mean MDE. This indicates a benefit of the transformer decoder block specifically

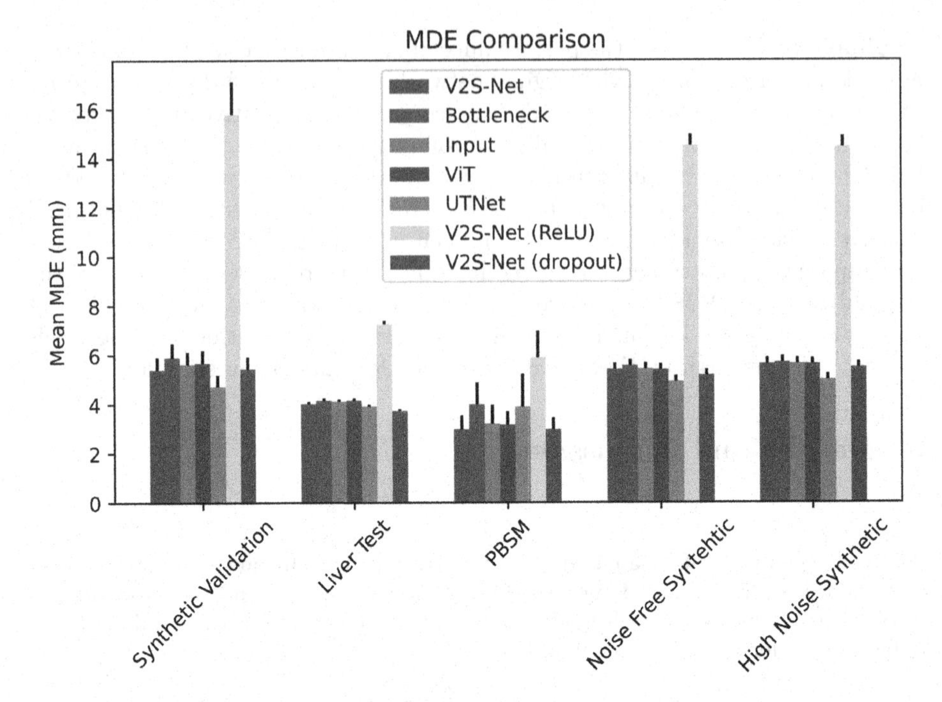

Fig. 4. Performance comparison between each network architecture under investigation and the baseline network architecture (V2S-Net) in terms of Mean Displacement Error - MDE (mm) evaluated across five datasets.

in terms of semantic encoding; this network block appears to be able to carry over global features in a manner that the simple skip connection cannot.

Results in terms of elementary modifications to the V2S-Net were similarly unremarkable. Unexpectedly, the changing of the activation function led to a substantial increase in MDE. It is possible that the nature of the output as a signed function creates issues when using the strictly non-negative ReLU function. Combined with the need of the network to output multiple resolution levels during training, this could reduce the ability of the network to generate effective estimates. On the other hand, the use of dropout has a more negligible effect on MDE. The current understanding of the dropout indicates that the training set is not too restricted to cause substantial network overfitting.

The need for substantial variability in input shape creates a demand for large quantities of synthetically-acquired data, as is the case in this study. Current work is investigating methods to generate novel liver meshes that are still physiologically plausible. It is important to note that the current analysis is specifically tested on the purpose of navigation in liver surgery. As such, it is not necessarily problematic if the method is overfitted to liver shapes, as long as it is generalized enough to adapt to novel liver shapes.

It may also be feasible to consider alternative methods for encoding of liver shapes. The current implementation with fixed inputs of $64 \times 64 \times 64$ voxels does require substantial computational power to increase the resolution; hence, further boosting the resolution will require techniques that provide a reasonable trade-off between resolution and computational expense. Prior work has identified deep networks trained on functional map representations as a viable method for non-rigid partial shape correspondence [3]. Modifications to that methodology may provide another efficient method for volume to surface registration through encoding at arbitrary resolution.

The use of voxelized datasets as input and output makes it difficult to compare the performance of the models with other benchmarks used for similar tasks. Several investigations are currently being conducted to effectively convert the voxelized displacement estimates into a displaced mesh. This conversion to a more traditional displacement dataset will facilitate the comparison of the proposed model performance to the performance of a broader set of existing techniques. Typical metrics used for assessing similar tasks have included the mean error value at mesh nodes as in Suwelack *et al.* [26]; and Hausdorff distance between the pre-operative and intraoperative meshes as in Elhawary *et al.* [9]. Future updates to this framework that can easily improve these metrics will allow for more unified comparison with traditional methods and benchmarks.

4 Conclusion and Future Work

In this work we investigate several network architecture modifications and extensions to baseline configurations featuring the classic U-Net architecture in the effort to improve the performance of voxelized volume-to-surface liver registration. This study has shown that, using synthetically generated data, the network configurations investigated here were able to predict displacement fields within 5 mm on average of the ground truth displacements. Moreover, while three of the transformer-based modifications did not yield significant performance improvements in terms of the quantified mean displacement error (MDE), the UTNet transformer modification led to the most significant performance improvement, while the dropout and ReLU activation functions led to slight and significant performance deterioration, respectively. Nevertheless, the UTNet-based transformer architecture not only improved overall performance, yielding a MDE on the order of 4 mm relative to the ground truth displacement, but also brings forth several advantages over other methods, specifically: it performs both a rigid and nonrigid registration concurrently, does not require any parameter tuning, and does not rely on any prior knowledge of boundary conditions.

Several avenues exist for further extensions of this work. Pfeiffer *et al.* [22] pointed out the potential of training the networks on inhomogeneous bodies to more accurately capture the nature of lesion-containing organs. This could allow for further extensions of the network by allowing for estimates of the ground truth material property to be passed in as input [22]. While exact knowledge of these properties is not available, reasonable estimates may suffice to solve the nonrigid registration. In addition, we also plan to extend the validation of the robustness of the best performing models using more realistic, either *in vitro* collected data or deidentified clinical patient data.

Acknowledgements. Research reported in this publication was supported by the National Institute of General Medical Sciences Award No. R35GM128877 of the National Institutes of Health, the Office of Advanced Cyber Infrastructure Award No. 1808530 of the National Science Foundation, and the Division Of Chemistry, Bioengineering, Environmental, and Transport Systems Award No. 2245152 of the National Science Foundation.

References

1. Acidi, B., Ghallab, M., Cotin, S., Vibert, E., Golse, N.: Augmented reality in liver surgery. J. Visceral Surg. **160**(2), 118–126 (2023)
2. Antonelli, M., et al.: The medical segmentation decathlon. Nat. Commun. **13**(1), 4128 (2022)
3. Attaiki, S., Pai, G., Ovsjanikov, M.: DPFM: deep partial functional maps. In: 2021 International Conference on 3D Vision (3DV), pp. 175–185 (2021)
4. Barcali, E., Iadanza, E., Manetti, L., Francia, P., Nardi, C., Bocchi, L.: Augmented reality in surgery: a scoping review. Appl. Sci. **12**(14), 6890 (2022)
5. Barequet, G., Sharir, M.: Partial surface and volume matching in three dimensions. IEEE Trans. Pattern Anal. Mach. Intell. **19**(9), 929–948 (1997)
6. Blender Online Community. Blender - a 3D modelling and rendering package. Blender Foundation, Blender Institute, Amsterdam
7. Corona-Figueroa, A., Frawley, J., Bond-Taylor, S., Bethapudi, S., Shum, H.P.H., Willcocks, C.G.: MedNeRF: medical neural radiance fields for reconstructing 3D-aware CT-projections from a single X-ray (2022)
8. Dosovitskiy, A., et al.: An image is worth 16×16 words: transformers for image recognition at scale (2021)
9. Elhawary, H., et al.: Multimodality Non-rigid image registration for planning, targeting and monitoring during CT-guided percutaneous liver tumor cryoablation. Acad. Radiol. **17**(11), 1334–1344 (2010)
10. Galle, P.R., et al.: EASL clinical practice guidelines: management of hepatocellular carcinoma. J. Hepatol. **69**(1), 182–236 (2018)
11. Gao, Y., Zhou, M., Metaxas, D.: UTNet: a hybrid transformer architecture for medical image segmentation (2021)
12. Gelfand, N., Mitra, N.J., Guibas, L.J., Pottmann, H.: Robust global registration (2005)
13. Geuzaine, C., Remacle, J.F.: Gmsh: a three-dimensional finite element mesh generator with built-in pre- and post-processing facilities. Int. J. Numer. Methods Eng. **79**, 1309–1331 (2009)
14. Hontani, H., Watanabe, W.: Point-based non-rigid surface registration with accuracy estimation. In: 2010 IEEE Computer Society Conference on Computer Vision and Pattern Recognition, pp. 446–452 (2010)
15. Li, H., Chen, H., Jing, W., Li, Y., Zheng, R.: 3D ultrasound spine imaging with application of neural radiance field method. In: 2021 IEEE International Ultrasonics Symposium (IUS), pp. 1–4 (2021)
16. Maki, H., Hasegawa, K.: Advances in the surgical treatment of liver cancer. BioSci. Trends **16**(3), 178–188 (2022)
17. Malinen, M., Råback, P.: Elmer finite element solver for multiphysics and multiscale problems. Multiscale Model. Methods Appl. Mater. Sci. **19**, 101–113 (2013)
18. Mendizabal, A., Márquez-Neila, P., Cotin, S.: Simulation of hyperelastic materials in real-time using deep learning. Med. Image Anal. **59**, 101569 (2020)

19. Mendizabal, Andrea, Tagliabue, Eleonora, Brunet, Jean-Nicolas., Dall'Alba, Diego, Fiorini, Paolo, Cotin, Stéphane.: Physics-based deep neural network for real-time lesion tracking in ultrasound-guided breast biopsy. In: Miller, Karol, Wittek, Adam, Joldes, Grand, Nash, Martyn P.., Nielsen, Poul M. F.. (eds.) MICCAI 2018-2019, pp. 33–45. Springer, Cham (2020). https://doi.org/10.1007/978-3-030-42428-2_4
20. Mildenhall, B., Srinivasan, P.P., Tancik, M., Barron, J.T., Ramamoorthi, R., Ng, R.: NeRF: representing scenes as neural radiance fields for view synthesis (2020)
21. Nakamura, K., et al.: The hepatic left lateral segment inverting method offering a wider operative field of view during laparoscopic proximal gastrectomy. J. Gastrointest. Surg. **24**(10), 2395–2403 (2020)
22. Pfeiffer, M., et al.: Non-rigid volume to surface registration using a data-driven biomechanical model (2020)
23. Rochester Institute of Technology. Research Computing Services (2019)
24. Ronneberger, Olaf, Fischer, Philipp, Brox, Thomas: U-net: convolutional networks for biomedical image segmentation. In: Navab, Nassir, Hornegger, Joachim, Wells, William M.., Frangi, Alejandro F.. (eds.) MICCAI 2015. LNCS, vol. 9351, pp. 234–241. Springer, Cham (2015). https://doi.org/10.1007/978-3-319-24574-4_28
25. Sivagami, S., Chitra, P., Kailash, G.S.R., Muralidharan, S.: UNet architecture based dental panoramic image segmentation. In: 2020 International Conference on Wireless Communications Signal Processing and Networking (WiSPNET), pp. 187–191 (2020)
26. Suwelack, S., et al.: Physics-based shape matching for intraoperative image guidance. Med. Phys. **41**(11), 111901 (2014)
27. Vaswani, A., et al.: Attention is all you need (2017)
28. Xiao, X., Guo, W., Chen, R., Hui, Y., Wang, J., Zhao, H.: A swin transformer-based encoding booster integrated in U-shaped network for building extraction. Remote Sens. **14**(11), 2611 (2022)
29. Yang, X., Kwitt, R., Niethammer, M.: Fast predictive image registration (2016)
30. Zeng, Y., Wang, C., Wang, Y., Gu, X., Samaras, D., Paragios, N.: Dense non-rigid surface registration using high-order graph matching. In: 2010 IEEE Computer Society Conference on Computer Vision and Pattern Recognition, pp. 382–389 (2010)

Vision Transformer-Based Self-supervised Learning for Ulcerative Colitis Grading in Colonoscopy

Ajay Pyatha[1], Ziang Xu[3], and Sharib Ali[2](✉)

[1] NepAl Applied Mathematics and Informatics Institute for Research (NAAMII), Kathmandu, Nepal

[2] School of Computing, University of Leeds, Leeds LS2 9JT, UK
s.s.ali@leeds.ac.uk

[3] Department of Engineering Science, University of Oxford, Oxford, UK

Abstract. Ulcerative colitis (UC) is a long-term condition that needs clinical attention and can be life-threatening. While Mayo Endoscopic Scoring is widely used to stratify patients at higher risk of developing colorectal cancer, the phenotypic endoscopic features involved in the scoring are highly inconsistent. Thus, devising automated methods is required. However, bias in the labels can also trigger such inconsistency and inaccuracy, which makes the use of fully supervised learning not preferable. We propose to exploit a self-supervised learning paradigm for automated MES grading of endoscopic images in UC. To take full advantage of local and global features, we propose to use Swin Transformers in the MoCo-v3 SSL setting. In addition, we provide a comprehensive benchmarking of other existing SSL methods. Our approach with Swin Transformer with MoCo-v3 provides performance boosts in different data size settings.

1 Introduction

Ulcerative colitis (UC) is a long-term condition characterised by inflammation of the bowel and rectum that can cause severe complications and thus require patient risk stratification and management. Due to the increased risk of cancer in patients with UC, a colonoscopy is usually performed together with a biopsy that further helps to stratify risk in patients affected with this disease. One of the most widely used scoring systems is the Mayo Endoscopic scoring [15] that has four categories ranging from 0 (normal or inactive condition); 1 (mild disease); 2 (moderate disease); and 3 (a severe disease with spontaneous bleeding and ulceration). These are based on endoscopic findings, such as erosions, vascular patterns, erythema, friability, and ulceration. Most of these phenotypic endoscopic features are further classified into mild, moderate, and severe, which makes it difficult for gastroenterologists to agree on their scoring. As a result, UC subjectivity [6] exists among the clinical experts, which needs to be addressed. Devising an automated deep-learning method can help reduce subjectivity. Fully

B. Bhattarai et al. (Eds.): DEMI 2023, LNCS 14314, pp. 102–110, 2023.
https://doi.org/10.1007/978-3-031-44992-5_10

supervised methods have been proposed in most literature for tackling UC grading [2,11,13,16]. We argue that due to the data-voracious nature of the supervised deep networks, a large number of expert-level annotations are required for incorporating a more extensive range of existing variabilities in these images, which makes them prone to being heavily biased towards that particular annotator and may not converge with other expert observations. Also, it is almost impossible to bring together multiple annotators to label, study and minimise biases in a massive pool of datasets. However, this can be more feasible when the data size is small.

Motivated by the idea of minimising variability in expert annotations, we aim to explore techniques that do not require many annotations but can be effectively trained using minor good quality labelled data providing higher accuracy. In this context, we are particularly interested in leveraging learnt feature embeddings using a representation learning technique that allows to encapsulate important phenotypic endoscopic features within the endoscopy data without requiring ground truth labels as a pre-text task which is also widely referred to as self-supervised learning (SSL). Such a setting can minimise the chances of subjectivity in ground truth labels as opposed to fully supervised learning techniques. The idea here is to fine-tune the model on a refined set of labelled data with higher confidence, allowing us to predict MES scores automatically and consistently. While the SSL technique using convolutional neural networks has been explored [17], to our knowledge, no method exploits vision transformers for UC classification. In addition, we provide a comprehensive benchmarking of other SSL methods with both CNN and transformer. Our approach with the Swin transformer with MoCo-v3 delivers a performance boost in different data size settings.

2 Related Work

2.1 UC Classification

Recent work by Polat *et al.* [14] involved assessing various supervised deep learning models on a large dataset of ulcerative colitis. Most of the works in the past have focused on supervised learning [1,11,16]. Only a few methods involve exploiting self-supervision strategy in this area [17]. Xu *et al.* [17,18] proposed patch-level instance-group discrimination with pretext-invariant representation learning and discrimination by clustering similar representative patches. However, they only used MoCo-v2 with a ResNet50 backbone in their work.

2.2 Self-supervised Learning

Self-supervised learning (SSL) uses pretext tasks to learn image representations from large amounts of unlabeled data and then fine-tune downstream tasks with small amounts of labelled data. Misra et al. [10] proposed pretext-invariant representation learning (PIRL), which uses Jigsaw puzzles to learn invariant representations. The SimCLR model was proposed by Chen et al. [3], using various data augmentation techniques. The contrastive loss is applied to maximize the inter-class similarity and minimize the same-class global and local similarity. He et al. [8] proposed the MoCo v1 model, which uses a contrastive loss to learn a general feature representation. MoCo v1 cleverly adds a dynamic memory queue to store feature vectors, significantly reducing the computational cost. Then He et al. [5] proposed MoCo v2 based on MoCo v1. Compared with MoCo v1, MoCo v2 uses more powerful data enhancement and cosine learning rate to improve the model effect further.

2.3 Vision Transformers

Due to the success of the Transformer in the field of natural language processing, Dosovitskiy et al. [7] applied Transformer to the area of computer vision and proposed the Vision Transformer model (ViT). ViT divides an image into fixed-sized patches through patch embedding, thus converting the visual problem into a sequence-to-sequence problem. The core conclusion of ViT is that when there is enough data for pre-training, ViT's performance will exceed that of CNN, breaking through the limitation of transformer lack of inductive bias, and can obtain a better migration effect in downstream tasks. On the self-supervised settings, while most previous approaches use traditional convolutional neural networks (CNNs), He et al. [4] proposed the MoCo v3 model utilising ViT as the backbone. In MoCo v3, since the Transformer has Q, K, and V structures with attention for long sequences, it can store and memorise more information, so the memory queue mechanism is cancelled and changed to use a large batch size for training. In order to solve the instability in training, MoCo v3 adopts the random patch projection method; that is, after random initialisation, the parameters of patch embedding are frozen. MoCo v3 achieves state-of-the-art results in each dataset. Later, three key features of a Swin Transformer network were introduced: inductive biases of locality, hierarchical feature representation, and translation invariance that makes them achieve a better speed-accuracy trade-off than other vision models.

3 Method

We propose to use a backbone feature encoding layer with Swin Transformer [9] for self-supervised learning using a framework MoCo v3 [4]. According to the setting of MoCo v3, we first take two crops for each endoscopic image and perform random data augmentation, which is then encoded by two Swin Transformer-based encoders providing query and key feature vectors f_q and f_k (Fig. 1). Positive and negative sets are built, say $k+$ and $k-$, where $k+$ refers to a group of all samples of the same image as the query and $k-$ is otherwise. A contrastive loss is minimised using InfoNCE [12]. It is noted that f_q has a backbone, the projection head, and an extra prediction head, while f_k only has a backbone and projection head.

For the feature encoding, we used the Swin-B base model with 128 channels in the hidden layers in the first stage [9]. We have used the default window size

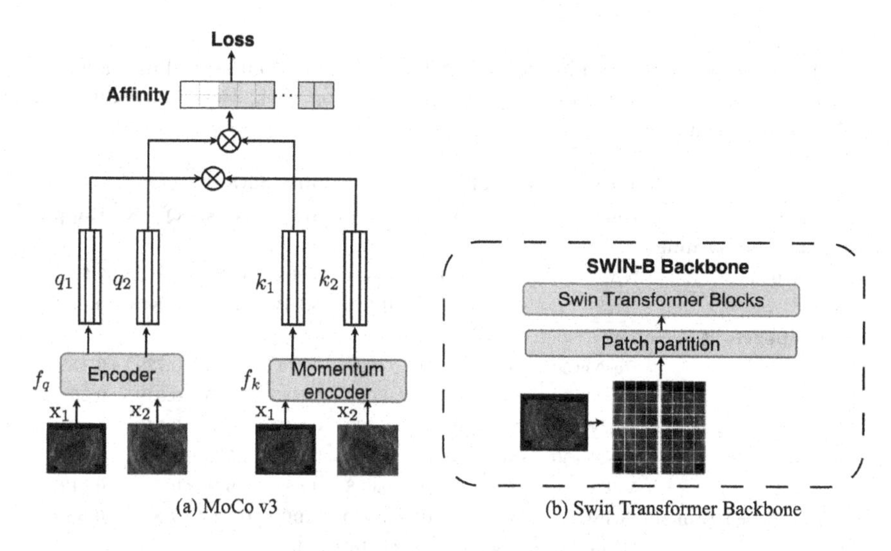

(a) MoCo v3 (b) Swin Transformer Backbone

Fig. 1. MoCo-v3 with SwinB for UC classification task.

of $M = 7$, the query dimension of each head to be $d = 32$ and the expansion layer of each MLP $\alpha = 4$. The choice of using the base model is because of the lower trainable parameters (88M), more than half than Swin-L (197M) but improved accuracy compared to Swin-T over the ImageNet classification task. Swin transformer uses patch partition, linear embeddings and two successive Swin Transformer blocks in the first stage, followed by patch merging and subsequent Swin Transformer blocks for the latter three additional stages. Both multi-head self-attention modules with regular and shifted windowing configurations are applied in each stage. An ImageNet pre-trained Swin-B extracts features for the feature encoding layers in the MoCo v3 SSL setting [4].

4 Experiments and Results

4.1 Dataset, Evaluation Metrics, and Experimental Setup

Dataset We have used Labeled Images for Ulcerative Colitis (LIMUC) consisting of a total of curated 11,276 endoscopic images from 1043 colonoscopy procedures of 564 UC patients at Marmara University Institute of Gastroenterology between December 2011 and July 2019 [14]. We have used official splits for all our experiments comprising a total of 9590 images for training and 15% of images (1686 images from 85 patients) for held-out samples for testing. We applied an 80:20 split for training and validation samples on the training set.

Table 1. Self-supervised learning approach for ulcerative colitis grading under 100% and 50% data availability for downstream classification task. Best two values are in bold for each setting.

Fine tuning with all available training data								
Method	**Backbone**	**Pre.**	**Top1**	**F1**	**Spec.**	**Sens.**	**QWK**	**Kappa**
Supervised learning								
ResNet50	NA	Yes	0.761	0.678	0.904	0.676	0.832	0.602
SwinB†	NA	Yes	**0.778**	**0.722**	**0.912**	**0.723**	**0.848**	**0.634**
Self-supervised learning								
PIRL	ResNet50	No	0.691	0.591	0.871	0.580	0.729	0.475
SimCLR	ResNet50	No	0.739	0.662	0.897	0.655	0.810	0.568
MoCo-v2	ResNet50	No	0.700	0.636	0.875	0.625	0.719	0.498
MoCo-v3	ResNet50	No	0.716	0.583	0.860	0.570	0.669	0.446
MoCo-v3	ViT	No	0.708	0.628	0.885	0.621	0.750	0.519
MoCo-v3 -SB(ours)	SwinB†	No	**0.767**	**0.711**	**0.905**	**0.717**	**0.844**	**0.613**
Fine tuning with 50% training data								
Method	**Backbone**	**Pre.**	**Top1**	**F1**	**Spec.**	**Sens.**	**QWK**	**Kappa**
Supervised learning								
ResNet50	NA	Yes	0.746	0.658	0.900	0.655	0.819	0.580
SwinB	NA	Yes	**0.767**	**0.672**	**0.909**	**0.675**	**0.846**	**0.614**
Self-supervised learning								
PIRL	ResNet50	No	0.648	0.592	0.847	0.513	0.600	0.388
SimCLR	ResNet50	No	0.719	0.641	0.890	0.633	0.803	0.536
MoCo-v2	ResNet50	No	0.657	0.566	0.856	0.555	0.626	0.422
MoCo-v3	ResNet50	No	0.676	0.586	0.863	0.646	0.659	0.449
MoCo-v3	ViT	No	0.708	0.630	0.882	0.616	0.736	0.515
MoCo-v3 -SB(ours)	SwinB†	No	**0.765**	**0.702**	**0.905**	**0.705**	**0.837**	**0.611**

NA, not available; † swin_base_patch4_window7_224 model; Pre., pretrained ImageNet weights; Spec., specificity; Sens., sensitivity

Table 2. Self-supervised learning approach for ulcerative colitis grading under 25% data availability for downstream classification task. Best values are in bold.

Method	**Backbone**	**Pre.**	**Top1**	**F1**	**Spec.**	**Sens.**	**QWK**	**kappa**
Supervised learning								
ResNet50	NA	Yes	0.729	0.643	0.900	0.665	0.801	0.563
SwinB	NA	Yes	0.724	0.620	0.888	0.602	0.748	0.537
Self-supervised learning								
PIRL	ResNet50	No	0.631	0.467	0.841	0.449	0.504	0.357
SimCLR	ResNet50	No	0.686	0.601	0.875	0.595	0.743	0.481
MoCo-v2	ResNet50	No	0.633	0.488	0.856	0.466	0.494	0.36
MoCo-v3	ResNet50	No	0.651	0.524	0.848	0.499	0.536	0.394
MoCo-v3	ViT	No	0.691	0.583	0.875	0.566	0.684	0.481
MoCo-v3 -SB(ours)	SwinB†	No	**0.765**	**0.700**	**0.903**	**0.693**	**0.831**	**0.606**

Table 3. Classification (top 1) accuracy per class for different sample sizes. Fine-tuning data using all, 50% of samples, and 25% of training samples.

Methods	Top 1 on full data				Top 1 on 50% data				Top 1 on 25% data			
	MES 0	MES 1	MES 2	MES 3	MES 0	MES 1	MES 2	MES 3	MES 0	MES 1	MES 2	MES 3
Supervised learning												
ResNet50	0.853	0.666	0.582	0.624	0.859	0.640	0.548	0.575	0.835	0.590	0.587	0.65
SwinB	0.868	**0.663**	**0.655**	0.708	**0.888**	**0.651**	**0.525**	0.633	0.851	**0.627**	0.514	0.416
Self-supervised learning												
PIRL (ResNet50)	0.860	0.487	0.458	0.517	0.854	0.413	0.277	0.508	0.83	0.493	0.192	0.283
SimCLR (ResNet50)	0.854	0.621	0.537	0.608	0.835	0.616	0.446	0.633	0.81	0.562	0.435	0.575
MoCo-v2 (ResNet50)	0.82	0.551	0.514	0.616	0.807	0.495	0.378	0.541	0.835	0.454	0.259	0.316
MoCo-v3 (ResNet50)	0.836	0.491	0.395	0.558	0.824	0.522	0.407	0.525	0.828	0.508	0.293	0.366
MoCo-v3 (ViT)	0.824	0.594	0.474	0.592	0.822	0.603	0.491	0.550	0.828	0.596	0.367	0.475
Ours	**0.883**	0.588	0.615	**0.783**	0.870	0.644	**0.525**	**0.783**	**0.878**	0.625	**0.593**	**0.675**

Evaluation metrics We have used standard top-k accuracy (percentage of samples predicted correctly), F1-score ($= \frac{2tp}{2tp+fp+fn}$, tp: true positive, fp: false positive), specificity ($= \frac{tp}{tp+fn}$) and sensitivity ($= \frac{tn}{tn+fp}$), quadratic weighted Kappa (QWK) and Cohen's Kappa for classification task of MES-scoring (0–3) for UC. The reason behind using QWK and Kappa is that there is a class imbalance in the dataset, and an ordinal relation exists between classes [14].

Experimental setup All models were trained on NVIDIA Tesla P100 GPU with a batch size of 32. The default learning rate of each self-supervised learning (SSL) model was used during the training of the pre-text task. We empirically set the learning rate for the downstream task to 10^{-4}. We used an Adam optimiser to minimise cross-entropy loss. All images were resized to 224×224. SSL approach with ResNet50 backbone took approximately 8 h to train, while those with vision transformers took around 48 h for training.

4.2 Results

Table 1 shows different data configurations for fine-tuning downstream classification tasks of the self-supervised learning (SSL) approaches. It also illustrates the fully supervised training of backbones used in the SSL approaches. It can be observed that utilising 100% training data Swin Transformers yielded the best result that reflects in the SSL setting (i.e., MoCo-V3 + SwinB backbone). An improvement over 8.3% compared to the proposed initial MoCo-v3 with ViT [4]. From Table 1 and Table 2, it can be observed that as the data samples decrease, fully supervised methods tend to show a drastic decrease in performance (1.4% on 50 % data and 7% on 25% data with SwinB) compared to our proposed SSL configuration (SwinB and MoCo-v3) that showed only negligible decrease. Our approach yielded consistent results across both 50% and 25% data with Top 1 accuracy of 76.5% in both cases. With 25% data, our method produced a 17.5% improvement over MoCo-v3 with ResNet50 backbone. Also, compared to the fully supervised model, our model outperformed SwinB by 5% on Top 1

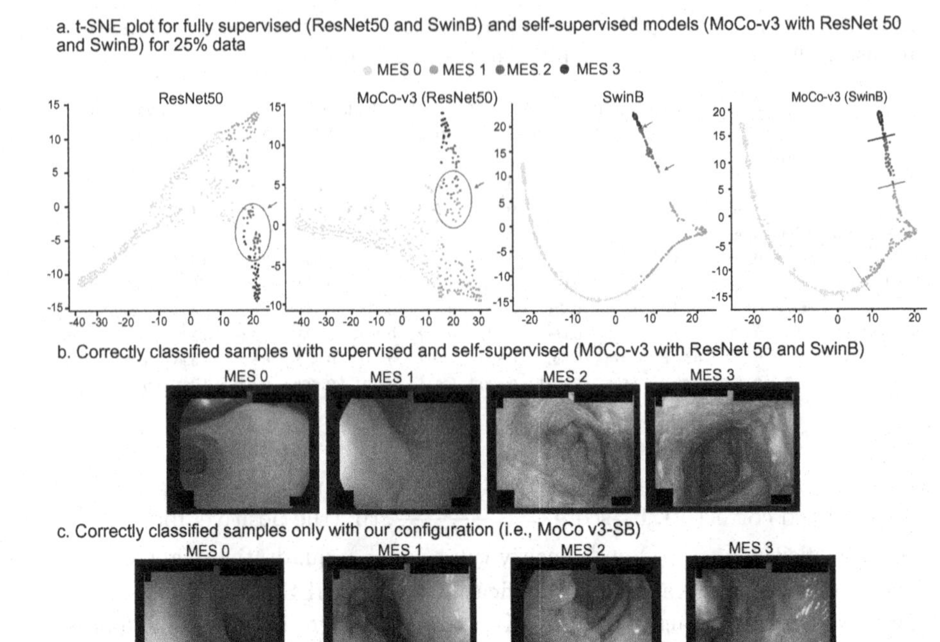

Fig. 2. UC classification on 25% data. a) Represents the 2D t-SNE plot showing confused classes in red ellipese and pointed with arrows. b) Samples that were correctly classified with methods indicated in (a). c) Illustration of samples that were misclassified by other approaches but only correctly classified by ours.

and 11% on the QWK metric. A similar trend can be observed across all other metrics.

A class-wise classification accuracy (top 1) is provided in Table 3. It can be observed that even on complete data, our approach outperformed all SSL approaches (with different backbones), including those with fully supervised settings. As going towards 25% data, we can observe that there is a consistent rise in the performance gap between the supervised backbone model SwinB and our approach with the same SwinB backbone in almost all classes (improvement of 3.1% for MES 0, nearly 8% for MES2, and 62% for MES 3).

It can be observed from Fig. 2 a that on 25% data, there exists a significant overlap between MES 2 and MES 3 and MES 0 and MES 1 in the supervised setting with ResNet50, which is lesser but still dominant in the SSL version with MoCo v3 with ResNet50 backbone (see red circles and arrows in Fig. 2a). An evident boundary separation can be observed with MoCo v3 with SwinB backbone (see straight lines). Similarly, from Fig. 2b, it is clear that most methods worked fine when the mucosa is distinctly clear with visible patterns in MES 2 and MES 3. However, the severity appearance of the MES is misjudged by other methods in Fig. 2c for all the MES scores.

5 Discussion and Conclusion

Ulcerative colitis (UC) is an inflammatory disease with relapses and remissions, and categorising different stages based on endoscopic appearances is challenging and often subjective. While efforts have been made to automate scoring systems in the past, these have been widely based on supervised learning paradigms known as data voracious, requiring labelled samples. This work addresses this problem, demonstrating that using a self-supervised learning paradigm with Swin Transformer-based backbone compared to traditional CNN-based models can provide competitive classification accuracies over different categories with only a few samples (Table 3). MoCo v3 with Swin-B backbone has lower class overlap (Fig. 2 a). This is because the Swin transformer-based encoder has hierarchical feature representation, which can capture very localised feature representation providing better inter-class separations. We also observed that due to the degree of severity and some phenotypic features (Fig. 2 c) are not captured by most methods where our proposed approach provides correct predictions. This could be because most of the areas in the endoscopic frames appear similar to other MES scores. For example, MES 1 is classified as MES 0 and MES 2 as MES 3, which can be because only very local features are distinguishable.

Thus, this work provides a benchmark for a self-supervised approach to ulcerative colitis. We demonstrate that using Swin Transformer-based backbone improves the classification and provides consistent results when the number of training samples is decreased from 100% to 50% to 25% during fine-tuning of the main classification task. In our future work, we aim to include other downstream tasks, such as detection and segmentation, on endoscopic and other medical imaging data.

References

1. Ali, S.: Where do we stand in ai for endoscopic image analysis? deciphering gaps and future directions. npj Digital Med. **5**(1), 184 (2022). https://doi.org/10.1038/s41746-022-00733-3
2. Becker, B.G., et al.: Training and deploying a deep learning model for endoscopic severity grading in ulcerative colitis using multicenter clinical trial data. Therapeutic advances in gastrointestinal endoscopy 14 (2021)
3. Chen, T., Kornblith, S., Norouzi, M., Hinton, G.: A simple framework for contrastive learning of visual representations. In: International Conference on Machine Learning, pp. 1597–1607. PMLR (2020)
4. Chen, X., Xie, S., He, K.: An empirical study of training self-supervised vision transformers. In: 2021 IEEE/CVF International Conference on Computer Vision (ICCV), pp. 9620–9629 (2021). https://doi.org/10.1109/ICCV48922.2021.00950
5. Chen, X., Fan, H., Girshick, R., He, K.: Improved baselines with momentum contrastive learning. arXiv preprint arXiv:2003.04297 (2020)

6. Cooney, R.M., Warren, B.F., Altman, D.G., Abreu, M.T., Travis, S.P.: Outcome measurement in clinical trials for ulcerative colitis: towards standardisation. Trials **8**(1), June 2007. https://doi.org/10.1186/1745-6215-8-17
7. Dosovitskiy, A., et al.: An image is worth 16 × 16 words: Transformers for image recognition at scale. arXiv preprint arXiv:2010.11929 (2020)
8. He, K., Fan, H., Wu, Y., Xie, S., Girshick, R.: Momentum contrast for unsupervised visual representation learning. In: Proceedings of the IEEE/CVF Conference on Computer Vision and Pattern Recognition, pp. 9729–9738 (2020)
9. Liu, Z., Lin, Y., Cao, Y., Hu, H., Wei, Y., Zhang, Z., Lin, S., Guo, B.: Swin transformer: hierarchical vision transformer using shifted windows. In: 2021 IEEE/CVF International Conference on Computer Vision (ICCV), pp. 9992–10002 (2021)
10. Misra, I., Maaten, L.v.d.: Self-supervised learning of pretext-invariant representations. In: Proceedings of the IEEE/CVF Conference on Computer Vision and Pattern Recognition, pp. 6707–6717 (2020)
11. Mokter, M.F., Oh, J., Tavanapong, W., Wong, J., Groen, P.C.d.: Classification of ulcerative colitis severity in colonoscopy videos using vascular pattern detection. In: International Workshop on Machine Learning in Medical Imaging, pp. 552–562. Springer (2020)
12. van den Oord, A., Li, Y., Vinyals, O.: Representation learning with contrastive predictive coding. ArXiv abs/1807.03748 (2018)
13. Ozawa, T., Ishihara, S., Fujishiro, M., Saito, H., Kumagai, Y., Shichijo, S., Aoyama, K., Tada, T.: Novel computer-assisted diagnosis system for endoscopic disease activity in patients with ulcerative colitis. Gastrointest. Endosc. **89**(2), 416–421 (2019)
14. Polat, G., Kani, H.T., Ergenc, I., Ozen Alahdab, Y., Temizel, A., Atug, O.: Improving the Computer-Aided Estimation of Ulcerative Colitis Severity According to Mayo Endoscopic Score by Using Regression-Based Deep Learning. Inflammatory Bowel Diseases p. izac226 (2022). https://doi.org/10.1093/ibd/izac226
15. Schroeder, K.W., Tremaine, W.J., Ilstrup, D.M.: Coated oral 5-aminosalicylic acid therapy for mildly to moderately active ulcerative colitis. N. Engl. J. Med. **317**(26), 1625–1629 (1987). https://doi.org/10.1056/NEJM198712243172603
16. Stidham, R.W., et al.: Performance of a deep learning model vs human reviewers in grading endoscopic disease severity of patients with ulcerative colitis. JAMA Netw. Open **2**(5), e193963–e193963 (2019)
17. Xu, Z., Ali, S., Gupta, S., Leedham, S., East, J.E., Rittscher, J.: Patch-level instance-group discrimination with pretext-invariant learning for colitis scoring. In: Machine Learning in Medical Imaging, pp. 101–110 (2022)
18. Xu, Z., Rittscher, J., Ali, S.: SSL-CPCD: self-supervised learning with composite pretext-class discrimination for improved generalisability in endoscopic image analysis. arXiv:2306.00197 (2023)

Task-Guided Domain Gap Reduction for Monocular Depth Prediction in Endoscopy

Anita Rau[1,2(✉)], Binod Bhattarai[1,3], Lourdes Agapito[1], and Danail Stoyanov[1]

[1] University College London, London, UK
[2] Stanford University, Stanford, USA
arau@stanford.edu
[3] University of Aberdeen, Aberdeen, UK

Abstract. Colorectal cancer remains one of the deadliest cancers in the world. In recent years computer-aided methods have aimed to enhance cancer screening and improve the quality and availability of colonoscopies by automatizing sub-tasks. One such task is predicting depth from monocular video frames, which can assist endoscopic navigation. As ground truth depth from standard in-vivo colonoscopy remains unobtainable due to hardware constraints, two approaches have aimed to circumvent the need for real training data: supervised methods trained on labeled synthetic data and self-supervised models trained on unlabeled real data. However, self-supervised methods depend on unreliable loss functions that struggle with edges, self-occlusion, and lighting inconsistency. Methods trained on synthetic data can provide accurate depth for synthetic geometries but do not use any geometric supervisory signal from real data and overfit to synthetic anatomies and properties. This work proposes a novel approach to leverage labeled synthetic and unlabeled real data. While previous domain adaptation methods indiscriminately enforce the distributions of both input data modalities to coincide, we focus on the end task, depth prediction, and translate only essential information between the input domains. Our approach results in more resilient and accurate depth maps of real colonoscopy sequences. The project is available here: https://github.com/anitarau/Domain-Gap-Reduction-Endoscopy.

Keywords: Depth prediction · Domain adaptation · Self-supervision · Endoscopy

1 Introduction

Colorectal Cancer is treatable if detected early, but patient outcome relies on the skill of the performing colonoscopist and complete diagnostic examination of the colon. To improve navigation during colonoscopy and assist endoscopists in

A. Rau and B. Bhattarai—Project conducted while at University College London.

B. Bhattarai et al. (Eds.): DEMI 2023, LNCS 14314, pp. 111–122, 2023.
https://doi.org/10.1007/978-3-031-44992-5_11

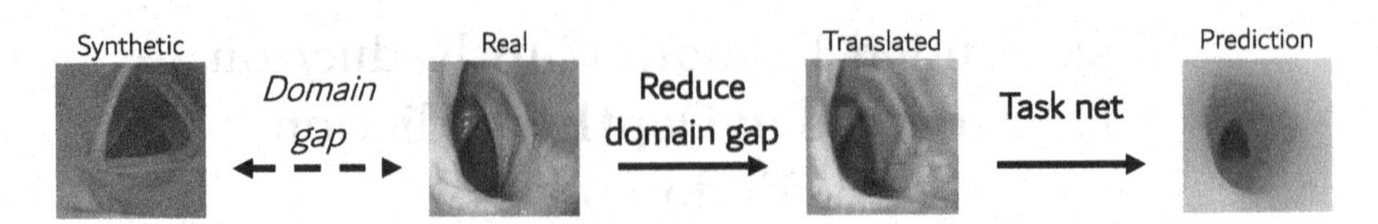

Fig. 1. The proposed network reduces the domain gap between synthetic and real images without fully closing it. We translate only domain- and task-specific information like water which is present in real images but not in synthetic ones.

ensuring complete examination, computer-assisted mapping and 3D reconstruction could help detect missed surfaces manifesting as holes in the colon map reconstruction [3,6]. Such surgical 3D environment maps could also be used for robotic systems and automation, but despite rapid advances in endoscopic artificial intelligence systems for polyp detection [1], mapping technologies remain challenging to implement robustly. Traditional methods require reliable features to be matched between frames, but colonoscopic images suffer from illumination inconsistency and a lack of texture. A featureless way to obtain a 3D model of the colon is to directly learn frame-wise depth and the relative camera pose between frames. But obtaining ground truth training data for real colonoscopy frames is currently unfeasible, as this would require a depth sensor to be integrated into a standard colonoscope. Instead, self-supervised methods [12,16] do not require any ground truth data and use warping-errors to optimize depth and pose predictions mutually. While such methods work well on homogeneous surfaces [17], they are challenged by the self-occluding tubular shape of the colon and the view-dependent illumination during colonoscopy.

An alternative to unlabeled real data is synthetically generated data with ground truth depth. Chen *et al.* propose to first train a network on synthetic data only and in a second, independent step, train the initialized network on real images with self-supervision [5]. However, this approach does not account for the domain shift between real and synthetic images. Other methods have used Generative Adversarial Networks (GANs) to reduce the appearance domain gap, some of which Fig. 2 depicts. Mahmood *et al.* [10] propose a multi-stage pipeline first mapping real examples to the synthetic domain, followed by an independent depth network trained on synthetic data only. However, independently training each sub-net might lead to sub-optimal results. Integrating domain adaptation and depth prediction into a mutual framework, Rau *et al.* [15] propose to train a single network on real and synthetic images. Mathew *et al.* propose a variation of a cycle GAN that maps virtual images to real images and vice versa [11]. One common drawback of these GAN-based methods is the holistic translation from one domain to another without considering domain- and task-specific components. Itoh *et al.* [8] are more deliberate about their choice of translation and decompose information based on a Lambertian-reflection mode; however, this hand-crafted decomposition is not guaranteed to extract and translate the most helpful information. All the translations mentioned above are difficult and distract from the main objective: predicting depth.

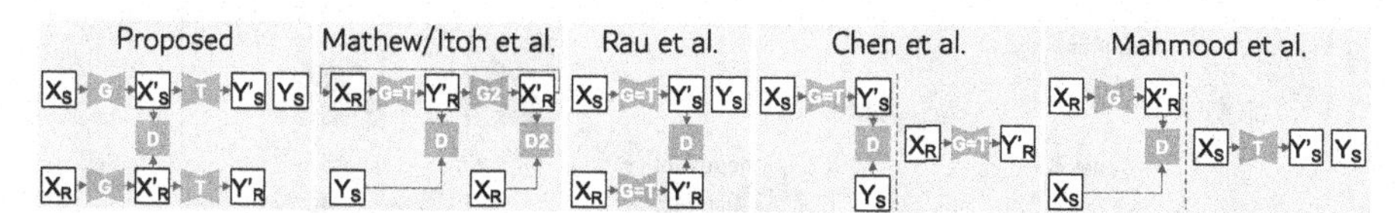

$X_{S/R}$ = Synthetic/Real image, X' = Generated image, $Y_{S/R}$ = Synthetic/Real depth, Y' = Generated depth. G = Generator, T = Task (depth) net, D = Discriminator.

Fig. 2. Comparisons of different domain adaptation methods [5,8,10,11,15] for depth prediction in colonoscopy. Depiction inspired by [19].

Rather than aligning one domain to another, images should reduce the domain gap between real and synthetic images only to the extent that it benefits the end task [13]. Our approach is end-to-end trainable and learns from real and synthetic data accounting for their different geometries by using separate depth losses. As Fig. 1 shows, the resulting network translates unknown geometric structures like water which is not present in the synthetic dataset. To the best of our knowledge, our method is the first to integrate synthetic data through a domain-adaptation framework into self-supervised depth estimation in colonoscopy.

2 Methods

A standard GAN-based approach to depth prediction can map real and synthetic images to the real domain [19], the synthetic/depth domain [15], or both [8,11]. To map an image $X_1 \in \mathcal{X}_1$ to a different domain $\mathcal{X}_2$, X_1 is passed through a generator $\mathcal{G}$. The output $\mathcal{G}(X_1)$ is then passed through the discriminator $\mathcal{D}$ which compares it to known images from $\mathcal{X}_2$. Minimizing

$$\mathbb{E}_{\mathcal{X}_2}[\log(\mathcal{D}(X_2))] + \mathbb{E}_{\mathcal{X}_1}[\log(1 - \mathcal{D}(\mathcal{G}(X_1)))], \tag{1}$$

forces $\mathcal{G}$ to learn the distribution of $\mathcal{X}_2$ [8,11,15,19].

There are two issues with this approach: (i) these GANs assume that real and synthetic depths come from the same distribution, which is not necessarily true; (ii) the domain adaptation is not guided by the end task. Depth losses for synthetic data are employed but there is no geometric supervision for predicted real depths. The domain adaptation network thus has no incentive to translate images such that the most accurate real depths are predicted. Our approach solves both issues.

2.1 Domain Gap Reduction

Our method maps as little information as possible to a mutual domain, allowing the network to focus on the end task. This concept was proposed by SharinGAN [13] for depth prediction from calibrated stereo cameras in urban settings.

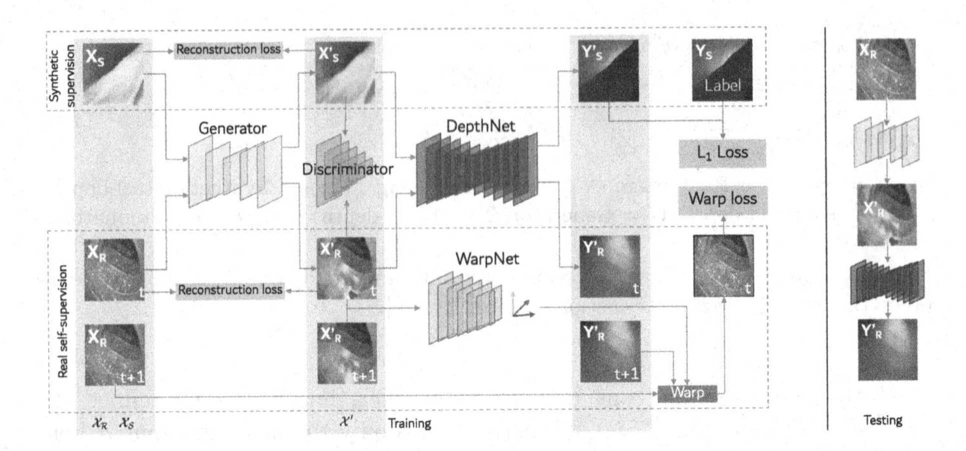

Fig. 3. Overview of our training and testing pipelines. Our network maps real and synthetic images into a new domain $\mathcal{X}'$. These translated images are passed through the DepthNet. Synthetic images are supervised with an L_1 error; real images are self-supervised using a warp loss. During testing, inputs are passed through the Generator and DepthNet only.

Figure 3 shows an overview of our approach. First, a GAN maps synthetic and real images to a mutual and end-task-specific domain. Being in the same domain, the synthetic and real images can now learn depth-specific features from one another. Let $\mathcal{X}_\mathcal{R}$ and $\mathcal{X}_\mathcal{S}$ denote the real and synthetic domain and let each image $\in \{\mathcal{X}_\mathcal{R}, \mathcal{X}_\mathcal{S}\}$ consist of domain-agnostic and domain-specific information.

Domain-agnostic information I is shared between the domains. Such information should encompass the underlying geometry of the colon. We actively avoid adaptation of I as it is unnecessary. Domain-specific information could be blood vessels that are visible in real images but not in synthetic ones. Such domain-specific information can be end-task-specific, δ_R and δ_S, or unspecific, δ'_R and δ'_S. The end task in our case is depth prediction but other tasks can be adapted to the same concept. Blood vessels do not encode relevant information about the geometry and are δ'_R; water and shadows, on the other hand, contain information about shape and are δ_S. The domain gap between δ'_R and δ'_S is negligible, as the depth net will learn to ignore such information. But the domain gap between δ_R and δ_S will affect the training of the depth network. If δ_R and δ_S are first mapped into a shared domain, real and synthetic data can complement each another.

Let x be a feature of image X. We want to learn a mapping $f : \mathcal{X}_\mathcal{R} \cup \mathcal{X}_\mathcal{S} \rightarrow \mathcal{X}'; x \mapsto f(x)$, such that $f(x) = x$ if $x \in \{\delta'_S, \delta'_R, \mathcal{I}\}$, and $f(x) \neq x$ if $x \in \{\delta_S, \delta_R\}$. But how do we learn such a mapping?

Instead of mapping one domain to the other, both domains can be mapped into a mutual one moving the means of the distributions together [2]:

$$L_{GAN} = \mathbb{E}_{\mathcal{X}_S}[\mathcal{D}(\mathcal{G}(X_S))] - \mathbb{E}_{\mathcal{X}_\mathcal{R}}[\mathcal{D}(\mathcal{G}(X_R))]. \quad (2)$$

To translate only crucial information in an image while retaining most of it, we use a reconstruction loss that penalizes translation by comparing the generator's input with its output:

$$L_R = \|\mathcal{G}(X_S) - X_S\|_2^2 + \|M(\mathcal{G}(X_R) - X_R)\|_1. \tag{3}$$

We experimentally found the $L1$-loss to lead to more similar reconstructions of small details in the real images and applied a specularity mask M based on the real images' RGB values.

Now, instead of having to learn how to translate a synthetic image to the real domain, or vice-versa, the network only needs to solve how to translate some information. To encourage that only task-relevant information is translated, we pass the generator's output through a depth net. The depth losses from *both* domains must then be back-propagated as described in the next section.

2.2 Depth Supervision

As labels for synthetic data exist, synthetic depths are supervised with an L_1-loss between the prediction and ground truth:

$$L_S = \|Y_S' - Y_S\|_1. \tag{4}$$

But as we miss ground truth for real data, the supervision for the real domain is less straight-forward. SharinGAN proposes to use stereo images for supervision. But in endoscopy we have to fall back to monocular video. We therefore propose to incorporate self-supervised geometric supervision for real images. Self-supervised losses help generalize to real anatomies but tend to converge to local minima. Additional synthetic supervision can help guide the optimization of self-supervised models.

For warping-based self-supervision we pass a second image, X^{t+1}, through the same generator and subsequently input both images into a WarpNet, which outputs a 6D pose vector $\mathbf{p}$ allowing us to warp image X^{t+1} to look like image X^t. We refer to this warped image as $X^{t+1 \to t}$. The warp loss L_W is computed as proposed in [12] allowing a direct comparison of both models:

$$L_W = 1 * L_{\text{photo}} + 0.5 * L_{\text{geo}} + 0.1 * L_{\text{smooth}}. \tag{5}$$

It consists of a photometric loss comparing an image to its warped counterpart:

$$L_{\text{photo}} = \sum \|\mathbf{T}(X^t) - X^{t+1 \to t}\|_2, \tag{6}$$

where $\mathbf{T}$ is the brightness-aware transformation of X according to [12]. Unlike [12] we do not incorporate the structural similarity index measure (SSIM) in the warp loss [18], making SSIM a fair evaluation measure on the test set. The geometric consistency loss is based on Y'^t warped to $t+1$, and Y'^{t+1} backwards interpolated to $\tilde{Y}'^{t+1}$ and the smooth loss supports convergence:

$$L_{\text{geo}} = \frac{\|Y'^{t\to t+1} - \tilde{Y}'^{t+1}\|_1}{Y'^{t\to t+1} + \tilde{Y}'^{t+1}}, \quad \text{and} \quad L_{\text{smooth}} = \sum(\exp^{-\nabla X^t} \cdot \nabla Y'^t)^2. \quad (7)$$

The final loss is a sum of the GAN-loss, reconstruction loss, and depth losses:

$$L = \omega_G L_{GAN} + \omega_R L_R + 0.5 \cdot (\omega_S L_S + \omega_W L_W). \quad (8)$$

Now that the task losses from both domains are back-propagated through the generator, the domain adaptation is guided by the end task and issue (ii) is addressed. Lastly, we observe that our network does not assume that real and synthetic depths are identically distributed (issue i); Fig. 2 illustrates that we only input RGB images (X_S, X_R) to a mutual discriminator, not depths.

2.3 Implementation Details

Our DepthNet is the architecture used both in EndoSLAM [12] and SharinGAN [13], allowing a direct comparison of the methods. For further comparability, we use the WarpNet proposed in [12]. We use SharinGAN's generator but replace the transposed convolutional layers with interpolation-based upsampling. We replace SharinGAN's discriminator with the lightweight discriminator proposed in Pix2Pix [7] reducing training time by almost half to 8 h on one NVIDIA A100-80GB GPU. The loss weights are chosen based on grid search and are: $\omega_G = 1, \omega_R = 10, \omega_S = 100, \omega_W = 1$. We train our network on 3,162 image pairs generated from 1,300 real colonoscopy frames of the EndoMapper[1] dataset [4]. All training images were extracted from a single video, as only two videos in the dataset provide camera intrinsics, and one was held out for testing. The synthetic dataset consists of 11,000 frames from the *Unity*-based SimCol[2] dataset [14].

3 Experiments

Evaluating a method that bypasses the need for training data is not straightforward, because the absence of test data is inherent to the task. Our evaluation thus first focuses on a qualitative comparison. We then quantitatively compare various reprojection losses for baselines trained in a self-supervised fashion. Lastly, we show that our method generalizes across patients and datasets.

Qualitative Comparison: We compare our method to two self-supervised approaches and two domain-adaptation-based algorithms. Our baselines are the self-supervised approaches EndoSLAM [12] and AF-SfMLearner [16] with all parameters set to their default values. The domain-adaptation-based baselines are: (1) a modification of SharinGAN [13], *SharinGAN**, in which we omit the

[1] https://www.synapse.org/Synapse:syn26707219/wiki/615178.

[2] https://www.ucl.ac.uk/interventional-surgical-sciences/simcol3d-data.

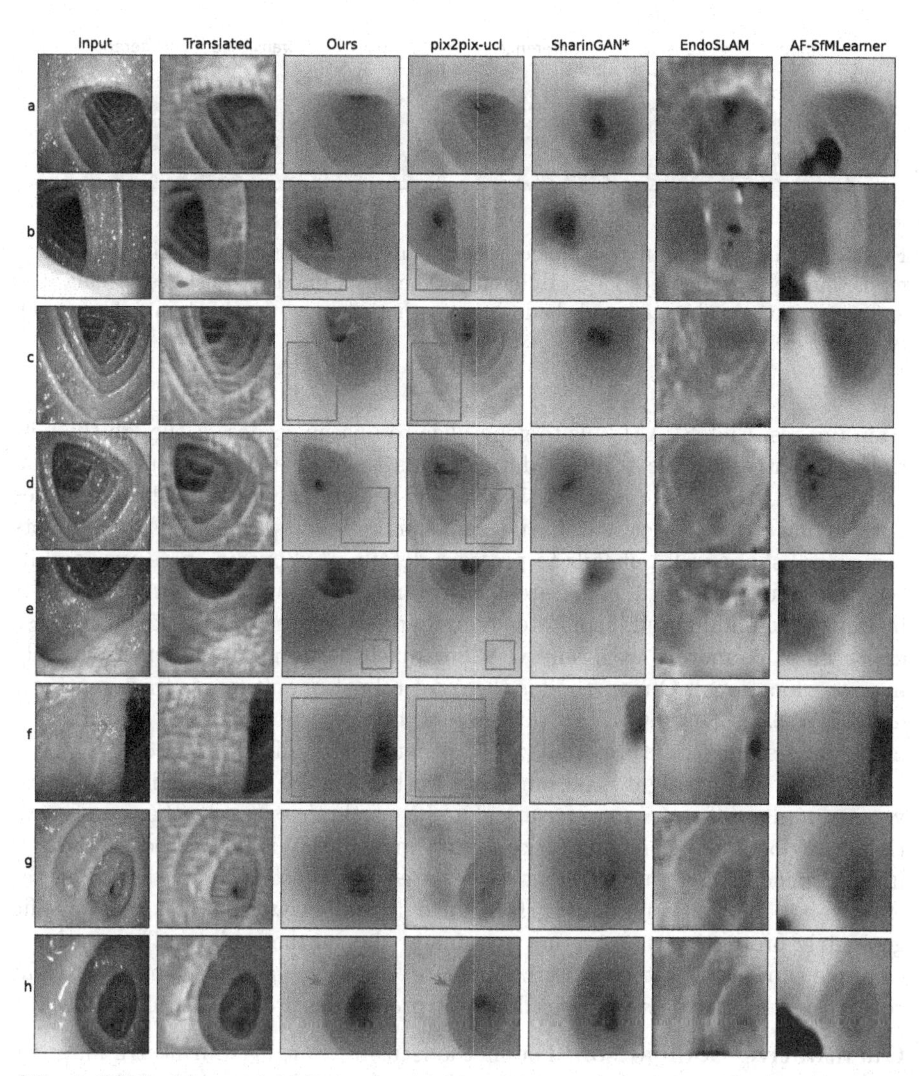

***SharinGAN without virtual supervision of real images as no stereo data is available.**

Fig. 4. Comparison of different methods on test images. EndoSLAM and the variation of SharinGAN fail to generalize to test data. AF-SfMLearner generalizes more robustly but suffers from large artefacts. We highlight some inconsistencies in pix2pix-ucl through magenta boxes. Our method is more resilient to specular highlights, water, and bubbles than the baselines and leads to smoother depth maps where the geometry is even (box in f) while preserving crisp edges (arrow in h) and details (arrows in c).

virtual supervision of the real images; (2) the extension of Pix2Pix [7], referred to as pix2pix-ucl [15]. Figure 4 depicts results on real test images. These test images are from the same patient but different sections of the colon than the

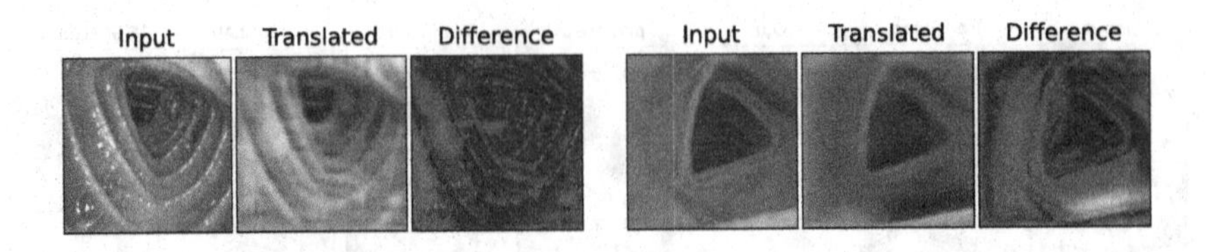

Fig. 5. Examples of input test images, their translated counterparts, and a difference map, where yellow denotes larger difference. The GAN mainly translates specularities and water in the real image, and shadows in synthetic images (see bottom right corner). (Color figure online)

train images. EndoSLAM fails to generalize to unseen scenes. Although the network converged on the training data, it fails to predict useful depth maps on test images. AF-SfMLeraner predicts largely sensible depth maps but struggles with artifacts like water, stool, and specular highlights. SharinGAN* predicts the overall shape well but fails to preserve details. Pix2pix-ucl, preserves details but the resulting depth maps are patchy and sometimes inaccurate. See, for instance, the highlighted inconsistency in map (e). Further, the gradient of these depth maps is uneven, with only very few pixels assuming high depths (mostly in wrong locations). Our method is the most robust one. It learns from synthetic images, such as SharinGAN or pix2pix-ucl, but incorporating the warping loss helps understand structures that would otherwise be misinterpreted.

In Fig. 5 we investigate how our GAN works. We plot an image, its translated version, and a difference map for a real and a synthetic example. We can observe that only domain-specific and task-relevant information is translated. In the real image, specularities and water are translated the most (yellow). Specularities encapsulate information about surface normals, while water puddles have specific geometric properties that are not present in the synthetic data. The synthetic example shows that the network hallucinates a strong shadow in the lower right corner, because *Unity's* renderer does not produce entirely realistic shadows.

Quantitative Comparison: For methods using warping during training, we can evaluate the warping losses on real test images. These indirectly give us information about the accuracy of the depth [5]. As our network is supervised by the warping loss *and* the synthetic L_1-loss, one might assume that our network

Table 1. Comparison of the different warping-based methods on 1,006 real test images.

	Photo. loss (Eq. 6) ↓	Geo. loss (Eq. 7) ↓	SSIM ↑
AF-SfMLearner	.096 ± .081 ‡	.069 ± .040	.686 ± .134 †
EndoSLAM	.076 ± .035 †	.061 ± .033 †	.641 ± .104 †
Proposed	.076 ± .036 †	.036 ± .031 †	.659 ± .110

†) Loss used for training as well. ‡) Related loss used for training as well.

results in a larger warp error than EndoSLAM, which is trained with warp loss only. However, in Table 1 we observe that our network results in a comparable photometric loss but a significantly smaller geometric loss. Comparing the SSIM between the methods, we observe that our model produces higher structural similarity than EndoSLAM, although EndoSLAM uses an SSIM-loss during training and our approach does not. A direct comparison of the self-supervised model EndoSLAM and our GAN suggests that synthetic data benefits our shared training approach. AF-SfMLearner produces the highest SSIM, though it was trained with the SSIM-loss while the proposed method was not. This evaluation has limitations. Warping errors evaluate the quality of depth and pose prediction mutually, and the two tasks can compensate each other. We also investigated EndoMapper's provided point clouds as potential pseudo ground truth but found them too noisy and sparse to be useful.

Generalization to New Patients: The EndoMapper dataset provides COLMAP results for two of the patients. These pseudo-labels are helpful as they provide camera intrinsics and because COLMAP rejects images that are too blurred or too occluded and thus neither useful to extract features nor for our purposes. We found that training on fewer but qualitatively better sequences improves results. We trained our model on one of the two patients with COLMAP labels and evaluated it on the other. Results are shown in Fig. 6. We can observe that the model generalizes well to a different patient, even when the colon is filled with water as in image (g, top row) or when geometries are peculiar as in

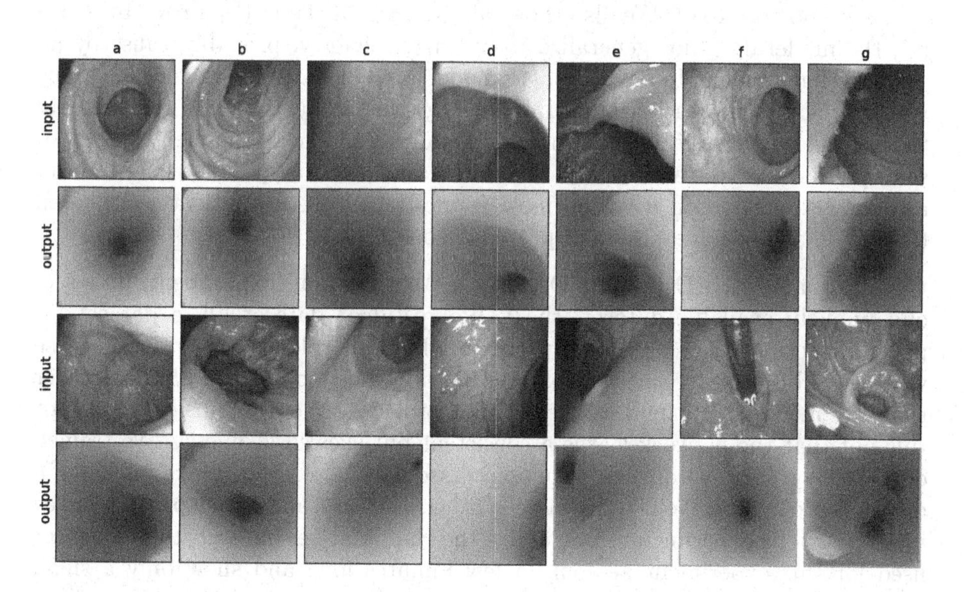

Fig. 6. Generalization to different patients. Our method predicts accurate depths even when trained on only one other patient. Failure cases due to extreme shade, tools, and very large specular highlights, that are not in the training set, are indicated by orange frames. (Color figure online)

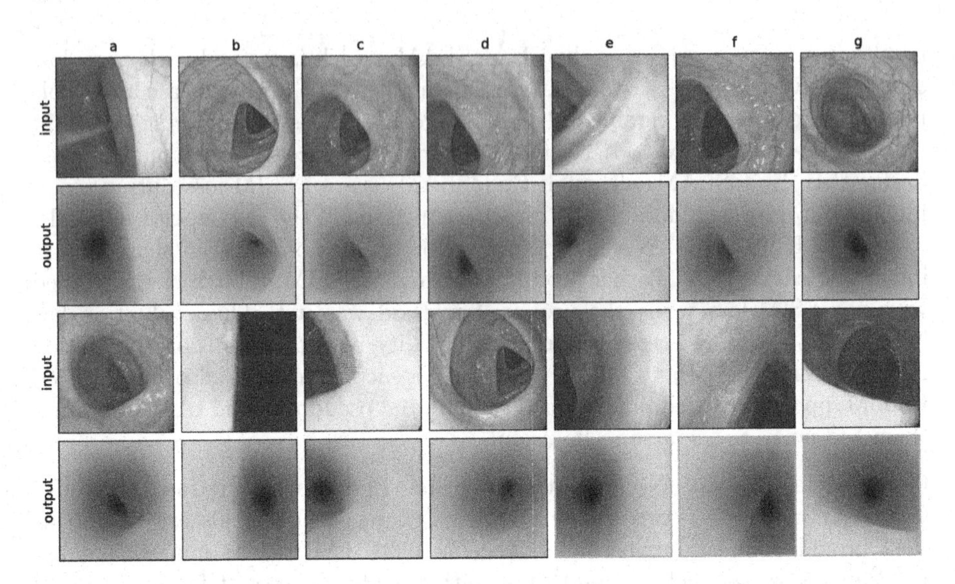

Fig. 7. Depth predictions on the LDPolypVideo dataset. Depicted are results on test images from a different procedure than the training data. Depths are accurate and robust to interlacing artefacts. Failure cases are indicated by orange frames. (Color figure online)

image (b, bottom row). We also show failure cases in the bottom row. In image (e), the model does not generalize to extensive shadow, probably cause by an occluded light source. In image (f), the model falsely locates the retroflexed scope viewing itself in the background. In image (g), the model does not generalize to the extraordinary large specular highlight. However, none of these extreme cases were present in the training data. Nonetheless, the model predicts sharp, accurate, and robust depth maps on most frames of an unseen patient, even when trained on a single anatomy and procedure only.

Generalization to Different Datasets: We repeat our experiments on a second publicly available dataset: the LDPolypVideo[3] dataset [9], a dataset for polyp detection that conveniently offers polyp-free colonoscopy videos. These videos can be used for our purposes as most frames focus on the lumen rather than the mucosa. As the dataset does not provide camera intrinsics we cannot rule out that the network learns a consistent but skewed geometry. We trained our model on frames from one colonoscopy sequence and applied it to images of a different procedure. We use the same synthetic dataset and hyper-parameters as in the previous experiments. But unlike the EndoMapper dataset, the sequences used for this experiment are only a few minutes long and show only a small section of a colon. Accordingly, the model is only trained on a fraction of the geometries observed in our first experiment. Nonetheless, the model can general-

[3] https://github.com/dashishi/LDPolypVideo-Benchmark.

ize to a different patient predicting accurate and sharp depth maps and is highly robust to interlacing artefacts as illustrated in Fig. 7.

4 Conclusions

Learning-based depth prediction has seen significant advances in recent years but requires labels for training, which are not available for colonoscopy. This work addresses the question how unlabeled real data and cheap, labeled synthetic data can be used in a mutual framework without overfitting to the geometry of the synthetic data. At the core of this work is the idea that domain adaptation is a challenging task that should only be addressed to the extent that it benefits the end task. Rather than indiscriminately translating entire images from one domain to another, and accounting only for appearance domain gaps, we propose task-guided domain gap reduction.

Our experiments show that our model learns to translate only task- and domain-specific information in real and synthetic input images. The network learns that water and air bubbles are specific to real data and that rendered shadows in synthetic data differ from real data. Accounting for these task-specific differences leads to geometrically consistent depth maps, outperforming previous domain translation and self-supervised models. We demonstrate that our results are more consistent with the smooth surfaces of the colon, more robust to unseen geometries, and still preserve details and edges. In the future, other tasks could benefit from task-guided domain gap reduction.

Acknowledgements. This work was supported by the Wellcome/EPSRC Centre for Interventional and Surgical Sciences (WEISS) [203145Z/16/Z]; Engineering and Physical Sciences Research Council (EPSRC) [EP/P027938/1, EP/ R004080/1, EP/P012841/1]; The Royal Academy of Engineering Chair in Emerging Technologies scheme; and the EndoMapper project by Horizon 2020 FET (GA 863146). All datasets used in this work are publicly available and linked in this manuscript. The code for this project is publicly available on Github. For the purpose of open access, the author has applied a CC BY public copyright license to any author-accepted manuscript version arising from this submission.

References

1. Ahmad, O.F., et al.: Establishing key research questions for the implementation of artificial intelligence in colonoscopy: a modified delphi method. Endoscopy **53**(09), 893–901 (2021)
2. Arjovsky, M., Chintala, S., Bottou, L.: Wasserstein generative adversarial networks. In: ICML, pp. 214–223. PMLR (2017)
3. Armin, M.A., et al.: Automated visibility map of the internal colon surface from colonoscopy video. IJCARS **11**(9), 1599–1610 (2016)
4. Azagra, P., et al.: Endomapper dataset of complete calibrated endoscopy procedures. arXiv preprint arXiv:2204.14240 (2022)
5. Cheng, K., et al.: Depth estimation for colonoscopy images with self-supervised learning from videos. In: MICCAI, pp. 119–128 (2021)

6. Freedman, D., et al.: Detecting deficient coverage in colonoscopies. IEEE TMI **39**(11), 3451–3462 (2020)
7. Isola, P., Zhu, J.Y., Zhou, T., Efros, A.A.: Image-to-image translation with conditional adversarial networks. In: CVPR., pp. 1125–1134 (2017)
8. Itoh, H., et al.: Unsupervised colonoscopic depth estimation by domain translations with a lambertian-reflection keeping auxiliary task. IJCARS **16**(6), 989–1001 (2021)
9. Ma, Y., et al.: Ldpolypvideo benchmark: a large-scale colonoscopy video dataset of diverse polyps. In: MICCAI, pp. 387–396 (2021)
10. Mahmood, F., Durr, N.J.: Deep learning and conditional random fields-based depth estimation and topographical reconstruction from conventional endoscopy. Med. Image Anal. **48**, 230–243 (2018)
11. Mathew, S., Nadeem, S., Kumari, S., Kaufman, A.: Augmenting colonoscopy using extended and directional cyclegan for lossy image translation. In: Proceedings of CVPR, pp. 4696–4705 (2020)
12. Ozyoruk, K.B., et al.: Endoslam dataset and an unsupervised monocular visual odometry and depth estimation approach for endoscopic videos. Med. Image Anal. **71**, 102058 (2021)
13. PNVR, K., Zhou, H., Jacobs, D.: Sharingan: combining synthetic and real data for unsupervised geometry estimation. In: Proceedings of CVPR, pp. 13974–13983 (2020)
14. Rau, A., Bhattarai, B., Agapito, L., Stoyanov, D.: Bimodal camera pose prediction for endoscopy. arXiv preprint arXiv:2204.04968 (2022)
15. Rau, A., et al.: Implicit domain adaptation with conditional generative adversarial networks for depth prediction in endoscopy. IJCARS **14**(7), 1167–1176 (2019)
16. Shao, S., et al.: Self-supervised monocular depth and ego-motion estimation in endoscopy: appearance flow to the rescue. Med. Image Anal. **77**, 102338 (2022)
17. Turan, M., et al.: Unsupervised odometry and depth learning for endoscopic capsule robots. In: IROS, pp. 1801–1807 (2018)
18. Wang, Z., Bovik, A.C., Sheikh, H.R., Simoncelli, E.P.: Image quality assessment: from error visibility to structural similarity. IEEE Trans. Image Process. **13**(4), 600–612 (2004)
19. Zheng, C., Cham, T.J., Cai, J.: T2net: synthetic-to-realistic translation for solving single-image depth estimation tasks. In: Proceedings of the European Conference on Computer Vision, pp. 767–783 (2018)

Author Index

B. Bhattarai et al. (Eds.): DEMI 2023, LNCS 14314, p. 123, 2023.
https://doi.org/10.1007/978-3-031-44992-5

Printed in the United States
by Baker & Taylor Publisher Services